Memoirs of the American Mathematical Society

Number 318

A. Baernstein II
and E. T. Sawyer

Embedding and multiplier theorems for $H^p(\mathbf{R}^n)$

Published by the

AMERICAN MATHEMATICAL SOCIETY

Providence, Rhode Island, USA

January 1985 · Volume 53 · Number 318 (end of volume)

MEMOIRS of the American Mathematical Society

SUBMISSION. This journal is designed particularly for long research papers (and groups of cognate papers) in pure and applied mathematics. The papers, in general, are longer than those in the TRANSACTIONS of the American Mathematical Society with which it shares an editorial committee. Mathematical papers intended for publication in the Memoirs should be addressed to one of the editors.

PREPARATION OF COPY. Memoirs are printed by photo-offset from camera-ready copy fully prepared by the authors. Prospective authors are encouraged to request a booklet giving detailed instructions regarding reproduction copy. Write to Editorial Office, American Mathematical Society, P.O. Box 6248, Providence, RI 02940. For general instructions, see last page of Memoir.

SUBSCRIPTION INFORMATION. The 1985 subscription begins with Number 314 and consists of six mailings, each containing one or more numbers. Subscription prices for 1985 are $188 list; $150 institutional member. A late charge of 10% of the subscription price will be imposed upon orders received from nonmembers after January 1 of the subscription year. Subscribers outside the United States and India must pay a postage surcharge of $10; subscribers in India must pay a postage surcharge of $15. Each number may be ordered separately; *please specify number* when ordering an individual number. For prices and titles of recently released numbers, refer to the New Publications sections of the NOTICES of the American Mathematical Society.

BACK NUMBER INFORMATION. For back issues see the AMS Catalogue of Publications.

Subscriptions and orders for publications of the American Mathematical Society should be addressed to American Mathematical Society, P.O. Box 1571, Annex Station, Providence, RI 02901-1571. *All orders must be accompanied by payment.* Other correspondence should be addressed to P.O. Box 6248, Providence, RI 02940.

MEMOIRS of the American Mathematical Society (ISSN 0065-9266) is published bimonthly (each volume consisting usually of more than one number) by the American Mathematical Society at 201 Charles Street, Providence, Rhode Island 02904. Second Class postage paid at Providence, Rhode Island 02940. Postmaster: Send address changes to Memoirs of the American Mathematical Society, American Mathematical Society, P.O. Box 6248, Providence, RI 02940.

The paper used in this journal is acid-free and falls within the guidelines established to ensure permanence and durability.

CONTENTS

ABSTRACT

The spaces $H^p(R^n)$, $0 < p \leq 1$, consist of tempered distributions f for which the maximal function $\sup_{t>0} |f*\psi_t(x)|$ belongs to $L^p(R^n)$. Here $\psi \in C_0^\infty$ with $\int \psi = 1$. We prove two main theorems. The first gives sharp conditions on the "size" of f which imply that f belongs to H^p. The conditions are phrased in terms of certain spaces K introduced by Herz. Our theorem may be regarded as the limiting endpoint version of a theorem by Taibleson and Weiss involving "molecules". We then use this embedding theorem to prove a sharp Fourier embedding theorem of Bernstein-Taibleson-Herz type.

Our other main theorem gives sharp sufficient conditions on $m \in L^\infty(R^n)$, for m to be a Fourier multiplier of H^p. This theorem also involves the K spaces and may be regarded as the limiting endpoint version of a multiplier theorem of Calderón and Torchinsky.

We also prove three results about Fourier transforms of H^p distributions. The first establishes the "lower majorant property" for H^p and the second is an $H^p(R^n)$ version of a recent theorem of Pigno and Smith about $H^1(\mathbb{T})$. The third result generalizes a theorem of Oberlin about growth of spherical means of $\hat{f}$, $f \in H^1(R^n)$.

__1980 Mathematics Subject Classification__ Primary: 42B30, 42B15.

Library of Congress Cataloging in Publication Data

Baernstein, Albert, 1941–

Embedding and multiplier theorems for $H^p(R^n)$

(Memoirs of the American Mathematical Society,
ISSN 0065-9266; 318 (Jan. 1985))

Bibliography: p.

1. Hardy spaces. 2. Embeddings (Mathematics) 3. Multipliers (Mathematical analysis) I. Sawyer, Eric T., 1951– . II. Title. III. Series: Memoirs of the American Mathematical Society; no. 318.

QA3.A57 no. 318 [QA331] 510s[515'.2433] 84-24294

ISBN 0-8218-2318-3

INTRODUCTION

The space $H^p(R^n)$, $n \geq 1$, $0 < p \leq 1$, consists of tempered distributions $f \in S'$ for which the maximal function

$$f^*(x) = \max_{t>0} |(f*\psi_t)(x)|, \qquad x \in R^n,$$

belongs to $L^p(R^n)$. Here ψ is any function in C_0^∞ with $\int_{R^n} \psi \, dx = 1$, and $\psi_t(x) = t^{-n}\psi(t^{-1}x)$. We define

$$\|f\|^p_{H^p} = \int_{R^n} (f^*)^p \, dx .$$

Many characterizations of $H^p(R^n)$ are given in [FS]. In particular, it is proved there that H^p is independent of the choice of ψ.

If f is a function on R^n which defines a tempered distribution one can ask what sorts of restrictions on the "size" of f will imply that $f \in H^p$. Taibleson and Weiss [TW] have found one such set of conditions. They call their functions "molecules". In §1 we present an embedding theorem of this sort which includes the Taibleson-Weiss results and is "sharp" in several respects. This enables us in §2 to prove sharp theorems of the form $\mathfrak{F} : X \to H^p$, where $\mathfrak{F}$ denotes Fourier transformation. These results are H^p analogues of theorems of S. Bernstein, Taibleson and Herz for L^p.

Received by the editors February 1, 1984.

The first author was supported by a grant from the National Science Foundation and the second author by a grant from the National Science and Engineering Research Council of Canada.

In §3 we formulate a Fourier multiplier theorem for H^p which sharpens up to their natural limits results of Calderón-Torchinsky [CT] and Taibleson-Weiss. In §4 we prove the embedding theorem and in §5 demonstrate its sharpness. §6 contains the proof of the multiplier theorem and §7 shows its sharpness.

Finally, in §8-10 we prove three theorems about Fourier transforms of H^p distributions which follow easily from the "atomic decomposition". The first asserts that H^p has the "lower majorant property" and answers a question of Weiss. The second contains an H^p analogue of a recent theorem of Pigno and Smith, while the last extends a theorem of D.M. Oberlin.

In some respects this paper may be regarded as a successor to [TW], and we are grateful to Professors Taibleson and Weiss for their friendly interest and encouragement. We also thank John Fournier for suggesting that we look in the direction of homogeneous Besov spaces in order to find sharp results. In [TW] the H^p distributions are defined as certain linear functionals on Lipschitz spaces. Latter's theorem about the atomic decomposition [L], see [Wi] for another proof and §6 of this paper for a description of the result, shows that the TW spaces $H^{p,\infty,N}$, where $N = [n(\frac{1}{p} - 1)]$, coincide with the H^p spaces as defined by us.

STANDING NOTATION

(0.1) N denotes the largest integer less than or equal to $n(\frac{1}{p} - 1)$.

(0.2) A_k denotes the shell $\{x \in R^n : 2^k \leq |x| \leq 2^{k+1}\}$, $k \in Z$.

(0.3) $H^p = H^p(R^n)$.

(0.4) $\int f = \int_{R^n} f(x)dx$, and dx denotes Lebesgue measure on R^n .

(0.5) C denotes a constant depending possibly on n and p which can change from line to line.

(0.6) For $x \in R^n$, x_i denote the coordinates of x .

(0.7) We use the usual multi-index notation. For $\beta = (\beta_1, \ldots, \beta_n)$ with each β_j a non-negative integer,

$$|\beta| = \sum_{j=1}^{n} \beta_j , \qquad x^\beta = x_1^{\beta_1} \cdots x_n^{\beta_n} .$$

(0.8) For $f \in C^N$, $P_N f$ denotes the N'th Taylor polynomial of f with basepoint $x = 0$.

1. EMBEDDING THEOREMS

If $f \in H^p$ then $|\hat{f}(\xi)| \leq C|\xi|^{n(\frac{1}{p}-1)}$ [TW, p. 105], so we expect that f will satisfy the vanishing moment condition

$$\int f(x)x^{\beta}\, dx = 0\ , \qquad 0 \leq |\beta| \leq N\ , \tag{1.1}$$

whenever the integrals make sense. Recall that N is the greatest integer less than or equal to $n(\frac{1}{p}-1)$ and integrals without limits are over all of R^n. Suppose that f does satisfy the necessary cancellation (1.1). What kind of size condition on f will guarantee that $f \in H^p$?

One such condition has been found by Taibleson and Weiss [TW, Theorem 2.9], who followed up on earlier work by Coifman and Weiss [CW]. Suppose that $0 < p \leq 1 \leq q \leq \infty$ and that $q > 1$ when $p = 1$. Taibleson and Weiss define a (p,q,N,ϵ) molecule centered at the origin to be a function $f \in L^q(R^n)$ which satisfies (1.1) and also

$$\int |f(x)|^q\, |x|^{\alpha q}\, dx < \infty\ ,$$

where α and ϵ are related by

$$\alpha = n(\frac{1}{p} - \frac{1}{q}) - n(\frac{1}{p} - 1) + \epsilon$$

and it is assumed that $\epsilon > n(\frac{1}{p} - 1)$. Thus $\alpha > n(\frac{1}{p} - \frac{1}{q})$. Theorem (2.9) of TW asserts that (p,q,N,ϵ) molecules belong to H^p.

We are going to prove a stronger embedding theorem which can be regarded as a critical endpointcase of the Taibleson-Weiss theorem. To formulate this theorem, and some others in this paper, we must introduce certain function spaces K and $\dot{K}$ which, as far as we know, were first considered by Herz [H]. Suppose that

$$1 \leq a \leq \infty, \qquad 0 \leq \alpha < \infty, \qquad 0 < b \leq \infty.$$

<u>Definition</u> (a) $\dot{K}_a^{\alpha,b}$ consists of all functions $f \in L^a_{loc}(\mathbb{R}^n \setminus \{0\})$ for which the norm or quasi-norm

$$(1.2) \qquad \|f\|_{\dot{K}_a^{\alpha,b}} = \{ \sum_{k=-\infty}^{\infty} (\int_{A_k} |f|^a)^{b/a} 2^{k\alpha b} \}^{1/b}$$

is finite.

(b) $K_a^{\alpha,b} = L^a \cap \dot{K}_a^{\alpha,b}$, with norm or quasi-norm

$$(1.3) \qquad \|f\|_{K_a^{\alpha,b}} = \|f\|_{L^a(\mathbb{R}^n)} + \|f\|_{\dot{K}_a^{\alpha,b}}.$$

Recall that $A_k = \{2^k \leq |x| \leq 2^{k+1}\}$. The usual modifications are made when $a = \infty$ or $b = \infty$.

The $\dot{K}$ spaces appear in [H], where they are denoted K^{α}_{ab}. Flett [F] gave a characterization of the Herz spaces which is easily seen to be equivalent to (1.2). They have been previously applied in H^p theory by Johnson [JO 2].

Elementary considerations show that the following inclusion relations are valid.

$$(1.4) \qquad \beta < \alpha \Rightarrow K_a^{\alpha,b} \subset K_a^{\beta,c},$$

$$(1.5) \qquad b \leq c \Rightarrow K_a^{\alpha,b} \subset K_a^{\alpha,c},$$

$$(1.6) \qquad a_1 \leq a_2 \Rightarrow K_{a_2}^{\alpha,b} \subset K_{a_1}^{\gamma,b}, \quad \text{where } \gamma = \alpha - n(\tfrac{1}{a_1} - \tfrac{1}{a_2})$$

Relations (1.5) and (1.6) are valid for the $\dot{K}$ spaces, but (1.4) is not.

The main case of interest to us occurs when $a = 1$, $\alpha = n(\frac{1}{p} - 1)$, $b = p$. Note that

$$\|f\|^p_{\dot{K}_1^{n(\frac{1}{p}-1),p}} = \sum_{k=-\infty}^{\infty} (\int_{A_k} |f|)^p \, 2^{kn(1-p)} ,$$

from which, for $p \leq 1$, follows

$$\|f\|_{L^p} \leq C \, \|f\|_{\dot{K}_1^{n(\frac{1}{p}-1),p}} \tag{1.7}$$

and

$$\int |f(x)| \, |x|^{n(\frac{1}{p}-1)} \, dx \leq C \, \|f\|_{\dot{K}_1^{n(\frac{1}{p}-1),p}} . \tag{1.8}$$

When $p < 1$ functions in $\dot{K}_1^{n(\frac{1}{p}-1),p}$ need not be locally integrable near the origin, and thus may not define distributions in S' in the usual way. However, if for $f \in \dot{K}_1^{n(\frac{1}{p}-1),p}$, $\varphi \in S$, and $N = [n(\frac{1}{p} - 1)]$ we define

$$\langle \varphi, f \rangle = \int (\varphi - P_N\varphi) f , \quad \text{when} \quad N < n(\tfrac{1}{p} - 1) , \tag{1.9}$$

$$\langle \varphi, f \rangle = \int (\varphi - P_{N-1}\varphi) f , \quad \text{when} \quad N = n(\tfrac{1}{p} - 1) \geq 1 \tag{1.10}$$

where $P_N\varphi$ denotes the N'th Taylor polynomial of φ at $x = 0$, then (1.8) shows that f defines a continuous linear functional on S', and hence for $0 < p < 1$ we may regard $\dot{K}_1^{n(\frac{1}{p}-1),p}$ as being continuously embedded in S' via the definitions (1.9) and (1.10).

Note that if $f \in \dot{K}_1^{n(\frac{1}{p}-1),p} \subset L^1$ and f satisfies the vanishing moment condition (1.1), then

$$\int f\varphi = \int f(\varphi - P_N\varphi) , \qquad \varphi \in S ,$$

so the distribution defined by f in the usual way coincides with the one defined by (1.9) and (1.10)

The Taibleson-Weiss (p,q,N,ϵ) molecules are just the functions in $K_q^{\alpha,q}$ for some $\alpha > n(\frac{1}{p} - \frac{1}{q})$ which satisfy (1.1). By (1.6), each of these molecules belongs to a space $K_1^{\alpha,q}$ for some $\alpha > n(\frac{1}{p} - 1)$.

Our sharp version of the Taibleson-Weiss theorem requires statement in three separate cases, according as $N < n(\frac{1}{p} - 1)$, $N = n(\frac{1}{p} - 1) \geq 1$, or $p = 1$. The theorems are proved in §4.

THEOREM 1a. *Suppose that* $N < n(\frac{1}{p} - 1)$. *Then* $\dot{K}_1^{n(\frac{1}{p} - 1),p} \subset H^p$ *and*

$$\|f\|_{H^p} \leq C\, \|f\|_{\dot{K}_1^{n(\frac{1}{p} - 1),p}} .$$

The theorem applies in particular to functions in $K_1^{n(\frac{1}{p} - 1),p}$ which satisfy (1.1). By (1.4) we have $K_1^{n(\frac{1}{p} - 1),p} \supset K_1^{\alpha,q}$ when $\alpha > n(\frac{1}{p} - 1)$, so this theorem generalizes TW for $0 < p < 1$, $N \neq n(\frac{1}{p} - 1)$. Note also that by stating the theorem for $\dot{K}$ we have dispensed with the global integrability hypothesis.

Theorem 1a is sharp in the sense that no larger K or $\dot{K}$ space is embedded in H^p. Choose $f_0 \in L^\infty$, $f_0 \not\equiv 0$, such that $\operatorname{supp} f_0 \subset A_0$ and, denoting surface area by $d\sigma$,

$$\int_{|x| = r} f_0(x) x^\beta \, d\sigma(x) = 0 \tag{1.11}$$

for every $|\beta| \leq N$ and $r \in (0,\infty)$. Define f by

$$f(x) = (2^{-nk} k^{-1})^{1/p} f_0(x 2^{-k}), \quad x \in A_k, \quad k \geq 0,$$

$$= 0, \quad |x| < 1 .$$

Then f satisfies (1.1) and belongs to $K_a^{n(\frac{1}{p}-\frac{1}{a}),q}$ for every $a \in [1,\infty]$ and $q > p$. Also $f \in L^1$ but $f \notin L^p$. Hence $f \notin H^p$.

Here is another way in which Theorem 1a is sharp. Let $\varepsilon(k)$, $k \geq 1$, be a given non-increasing sequence with $\lim_{k\to\infty} \varepsilon(k) = 0$. Then there is an $f \in L^1$ satisfying (1.1), $f = 0$ in $|x| < 1$, with

$$\sum_1^\infty \left(\int_{A_k} |f|\right)^p 2^{kn(1-p)} \varepsilon(k) < \infty ,$$

but $f \notin H^p$. To obtain such an f, select a subsequence $S \subset Z^+$ for which $\sum_{k\in S} \varepsilon(k) < \infty$. The desired function is given by

$$f(x) = 2^{-nk/p} f_0(x2^{-k}), \quad x \in A_k, \quad k \in S,$$

$$= 0, \quad \text{otherwise}.$$

There exists also a compactly supported example of this type. Define

$$f(x) = 2^{-nk/p} f_0(x2^{-k}), \quad x \in A_k, \quad k \in -S,$$

$$= 0, \quad \text{otherwise.}$$

Then

$$\sum_{-\infty}^{-1} \left(\int_{A_k} |f|\right)^p 2^{nk(1-p)} \varepsilon(-k) < \infty .$$

Take $\varphi \in C_0^\infty$ with compact support and $\int \varphi = 1$, and define $\varphi_t * f$ via (1.9). From (1.11) it follows that $\lim_{t\to 0} (\varphi_t * f)(x) = f(x)$, $x \neq 0$. Since $f \notin L^p$, it follows that $\sup_{t>0} |\varphi_t * f| \notin L^p$, and hence $f \notin H^p$.

In §5 we construct examples f_1 which show that when $N = n(\frac{1}{p} - 1)$ $\dot{K}_1^{n(\frac{1}{p}-1),q}$ is not contained in H^p, no matter how small q is. To state a replacement for Theorem 1a we introduce subspaces $Y(p) \subset \dot{K}^{n(\frac{1}{p}-1),p}$. Assume that $N = n(\frac{1}{p} - 1)$.

<u>Definition</u> $f \in Y(p)$ if

(1.12) $$\|f\|_{Y(p)} \equiv \sum_{k=-\infty}^{\infty} (\int_{A_k} |f|)^p \, 2^{kn(1-p)} (1+|k|) < \infty$$

and

(1.13) $$\int f(x) x^\beta \, dx = 0 \, , \quad \text{for every } |\beta| = N \, .$$

The integrals in (1.13) are absolutely convergent, by (1.8). For $f \in Y(p)$ the embeddings (1.9) and (1.10) of f in S' agree.

THEOREM 1b. <u>Suppose that</u> $N = n(\frac{1}{p} - 1)$ <u>and</u> $p < 1$. <u>Then</u> $Y(p) \subset H^p$ <u>and</u>

$$\|f\|_{H^p} \leq C \, \|f\|_{Y(p)} \, .$$

The theorem applies in particular when $f \in Y(p) \cap L^1$ and f satisfies (1.1). It's easy to show that $K_1^{\alpha,q} \subset Y(p)$ when $\alpha > n(\frac{1}{p} - 1)$, so this theorem generalizes TW for $0 < p < 1$, $N = n(\frac{1}{p} - 1)$. The examples f_1 and f_2 in §5 show that the weights $1 + |k|$ in (1.12) cannot be replaced by essentially smaller ones.

Theorem 1b becomes false when $p = 1$. In place of $Y(1)$ we have to consider a smaller space Y^*.

<u>Definition</u> $f \in Y^*$ if $f \in L'$, $\int f = 0$, and

(1.14) $$\|f\|_{Y^*} = \int |f(x)| \, (1 + \log^+ |\frac{f(x)}{\|f\|_{L^1}}| + \log^+ |x|) \, dx < \infty \, .$$

THEOREM 1c. $Y^* \subset H^1$ <u>and</u> $\|f\|_{H^1} \leq C \, \|f\|_{Y^*}$.

This is the $H^1(R^n)$ version of Zygmund's $L \log L$ theorem on the circle. We are surprised that no one seems to have discovered it before. It includes the Taibleson-Weiss theorem for $p = 1$, since $K_q^{\alpha,q} \subset Y^*$ for $q > 1$ and $\alpha > n(1 - \frac{1}{q})$. The examples f_1 in §5 show that $\log^+ |x|$ cannot be replaced by any essentially smaller weight, and the examples f_3 show that $L \log L$ cannot be replaced by any essentially weaker integrability condition.

Theorems 1a,b are about functions with possible non-integrable singularity at the origin. By translation invariance, the conclusions still hold if the rings A_k are replaced by rings $2^k \leq |x - x_0| \leq 2^{k+1}$. It would be interesting to find embedding theorems in which the one point singular set $\{x_0\}$ is replaced by a substantially larger one.

2. FOURIER EMBEDDING

In this section we use the embedding theorems of §1 to prove sharp H^p analogues of S. Bernstein's theorem which asserts that functions of class $\mathrm{Lip}\,\alpha$ on the circle have absolutely convergent Fourier series provided $\alpha > \frac{1}{2}$.

The Lipschitz or Besov spaces $B_a^{\alpha,b}$ are defined in Stein's book [S] to be the set of all $F \in L^a(\mathbb{R}^n)$ for which the norm or quasi-norm

$$(2.1)\qquad \|F\|_{B_a^{\alpha,b}} = \|F\|_{L^a(\mathbb{R}^n)} + \left(\int_0^\infty \left[t^{s-\alpha}\left(\int_{\mathbb{R}^n} \left|\frac{\partial^s u}{\partial t^s}(\xi,t)\right|^a d\xi\right)^{1/a}\right]^b \frac{dt}{t}\right)^{1/b}$$

is finite. Here $1 \le a \le \infty$, $\alpha \ge 0$, $0 < b \le \infty$, s denotes an integer larger than α, and $u(x,t)$ is the Poisson extension of f to $\mathbb{R}^n \times [0,\infty)$. Obvious modifications are made when $a = \infty$ or $b = \infty$. A brief account of the B spaces appear in [S, Chapter 5] where they are denoted $\Lambda_\alpha^{a,b}$. More thorough treatments have been given by Taibleson [T] who denotes them $\Lambda(\alpha\,; a,b)$, and by Peetre [P]. Stein and Taibleson consider only $b \ge 1$, but the results of interest to us are still valid for $0 < b < 1$.

Let $\mathfrak{F}$ denote the Fourier transformation.

TAIBLESON'S THEOREM [T, III]. *$\mathfrak{F}$ maps $B_2^{n(\frac{1}{p}-\frac{1}{2}),p}$ continuously into L^p, $0 < p \le 2$.*

Taibleson stated only the case $1 \le p \le 2$, but his argument works just as well when $0 < p \le 2$. The case $p = 1$ may be regarded as an n-dimensional non-periodic analogue of Bernstein's theorem. L^p theorems of this type in the periodic case had been proved earlier by Szasz [Sz] and Minakshisundaram and Szasz [MS].

Recall the spaces $K_a^{\alpha,b}$ introduced in §1. A more precise version of Taibleson's theorem is

(2.2) $\mathfrak{F}$ *maps* $B_2^{\alpha,q}$ *isomorphically onto* $K_2^{\alpha,q}$, $\alpha \geq 0$, $0 < q \leq \infty$.

To see that (2.2) implies Taibleson's theorem one uses the inclusion $K_2^{n(\frac{1}{p}-\frac{1}{2}),p} \subset L^p$, $0 < p \leq 2$, which is easily proved from the definition of the K-spaces. For the reader's convenience we will indicate at the end of this section how (2.2) follows from the definition (2.1).

The "homogeneous" version of (2.2), $\mathfrak{F} : \dot{B}_2^{\alpha,q} \to \dot{K}_2^{\alpha,q}$ isomorphically, is due to Herz [H]. This result furnishes the motivation for Peetre's definition of the Besov spaces in [P].

Returning to the non-homogeneous case, by (1.6) we have

$$(2.3) \qquad K_2^{n(\frac{1}{p}-\frac{1}{2}),p} \subset K_1^{n(\frac{1}{p}-1),p}, \qquad 0 < p \leq 1 .$$

From (1.8) if follows that

$$\int |f(x)|(1+|x|^{n(\frac{1}{p}-1)})dx \leq C\|f\|_{K_1^{n(\frac{1}{p}-1),p}} .$$

Hence (2.2) and (2.3) imply that $B_2^{n(\frac{1}{p}-\frac{1}{2}),p} \subset C^N$, where $N = [n(\frac{1}{p}-1)]$, and it makes sense to define

$$(2.4) \qquad X_2^{n(\frac{1}{p}-\frac{1}{2}),p} = \{F \in B_2^{n(\frac{1}{p}-\frac{1}{2}),p} : (D^\beta F)(0) = 0, \quad 0 \leq |\beta| \leq N\} .$$

THEOREM 2a. *If* $n(\frac{1}{p}-1) \notin Z$, $0 < p < 1$, *then* $\mathfrak{F}$ *maps* $X_2^{n(\frac{1}{p}-\frac{1}{2}),p}$ *continuously into* H^p.

PROOF. This follows at once from (2.2), (2.3), and Theorem 1a.

The examples following the statement of Theorem 1a show that $X_2^{n(\frac{1}{p}-\frac{1}{2}),p}$ cannot be replaced by any larger X space, for instance $X_2^{n(\frac{1}{p}-\frac{1}{2}),q}$ with $q>p$, and the examples f_1 in §5 show that when $n(\frac{1}{p}-1)\in\mathbb{Z}$ $X_2^{n(\frac{1}{p}-\frac{1}{2}),q}$ does not transform into H^p, no matter how small q is.

Suppose next that

$$\alpha>n(\frac{1}{p}-\frac{1}{2}), \quad 0<q\leq\infty .$$

Then (1.6) and (1.4) show that

$$K_2^{\alpha,q}\subset K_1^{\alpha-\frac{n}{2},q}\subset K_1^{n(\frac{1}{p}-1),p}, \tag{2.5}$$

and we may define $X_2^{\alpha,q}$ as in (2.4) with $B_2^{\alpha,q}$ in place of $B_2^{n(\frac{1}{p}-\frac{1}{2}),p}$.

THEOREM 2b. <u>If</u> $0<p\leq 1$, $0<q\leq\infty$ <u>and</u> $\alpha>n(\frac{1}{p}-\frac{1}{2})$ <u>then</u> $\mathfrak{F}$ <u>maps</u> $X_2^{\alpha,q}$ <u>continuously into</u> H^p.

PROOF. When $n(\frac{1}{p}-1)\notin\mathbb{Z}$ this follows from (2.2), (2.5) and Theorem 2a. If $n(\frac{1}{p}-1)\in\mathbb{Z}$ then $K_2^{\alpha,q}\subset Y(p)$ for $p<1$, $K_2^{\alpha,q}\subset Y^*$ for $p=1$, so the result follows from (2.2) and Theorems 1b,c.

A particularly interesting case of Theorem 2b is obtained when $q=2$. From (2.2) it follows that

$$\|F\|^2_{B_2^{\alpha,2}}\approx\int|\hat{F}(x)|^2\,(1+|x|^{2\alpha})\,dx .$$

Moreover, if $\alpha>n(\frac{1}{p}-\frac{1}{2})$ and $F\in B_2^{\alpha,2}$ then the condition $\int|F(\xi)|^2\,|\xi|^{-2\alpha}\,d\xi<\infty$ implies $F\in X_2^{\alpha,2}$. Consequently, we obtain the following simple sufficient condition for $\hat{F}$ to belong to H^p.

COROLLARY 1. Suppose that $\alpha > n(\frac{1}{p} - \frac{1}{2})$, $0 < p \le 1$, and that

$$\Phi_\alpha^2(F) \equiv \int (|F(\xi)|^2 + |F(\xi)|^2 |\xi|^{-2\alpha} + |\hat{F}(\xi)|^2 |\xi|^{2\alpha}) d\xi < \infty .$$

Then $\hat{F} \in H^p$ and $\|\hat{F}\|_{H^p} \le C\, \Phi_\alpha(F)$.

It is also possible to prove an H^p Fourier embedding theorem for the homogeneous Besov spaces $\dot{B}_a^{\alpha,b}$, denoted $\Lambda_{a,b}^{\alpha}$ by Herz, who proved the analogue of (2.2) for $\dot{B}$ and $\dot{K}$. Johnson [JO 1] gave an alternate approach to the $\dot{B}$ spaces and Flett [F] gave another proof of Herz's theorem. Janson [Ja] has recently written an interesting paper on this subject.

Suppose that $0 < p < 1$ and $n(\frac{1}{p} - 1) \notin Z$. The elements of $\dot{B}_2^{n(\frac{1}{p} - \frac{1}{2}),p}$ are equivalence classes of tempered distributions modulo polynomials of degree $\le N$. [H, p. 289]. By [H, p. 313] we have

$$\dot{B}_2^{n(\frac{1}{p} - \frac{1}{2}),p} \subset \dot{B}_\infty^{n(\frac{1}{p} - 1),p} \subset \dot{B}_\infty^{n(\frac{1}{p} - 1),\infty} ,$$

and the representing distributions in this last space are in fact C^N functions whose derivatives of order N satisfy a Lipschitz condition of order $n(\frac{1}{p} - 1) - N$. So, each equivalence class in $\dot{B}_2^{n(\frac{1}{p} - \frac{1}{2}),p}$ contains a unique representative $G \in C^N$ which satisfies $(D^\beta G)(0) = 0$, $|\beta| \le N$. It follows from Taylor's integral formula that

$$|G(\xi)| \le C\, |\xi|^{n(\frac{1}{p} - 1)} . \tag{2.6}$$

THEOREM 2c. <u>If</u> $0 < p < 1$, <u>and</u> $n(\frac{1}{p} - 1) \notin Z$ <u>then each equivalence class in</u> $\dot{B}_2^{n(\frac{1}{p} - \frac{1}{2}),p}$ <u>contains a unique function</u> G <u>for which</u> $\hat{G} \in H^p$. <u>Moreover</u>

$$\|\hat{G}\|_{H^p} \leq C \|G\|_{\dot{B}_2^{n(\frac{1}{p} - \frac{1}{2}),p}} .$$

PROOF. Consider the function G satisfying (2.6). Then $\hat{G} \in S'$. Herz's Bernstein-type theorem [H, Theorem 1], or [F, p. 547], asserts the existence of $g \in \dot{K}_2^{n(\frac{1}{p} - \frac{1}{2}),p}$ for which

$$\langle \check{G}, \varphi \rangle = \int g \varphi \tag{2.7}$$

for every $\varphi \in S$ which vanishes in a neighborhood of the origin. Define $\hat{g}_1 \in S'$ by

$$\langle g_1, \varphi \rangle = \int g(\varphi - P_N\varphi) , \quad \varphi \in S . \tag{2.8}$$

Since $\dot{K}_2^{n(\frac{1}{p} - \frac{1}{2}),p} \subset \dot{K}_1^{n(\frac{1}{p} - 1),p}$ by (1.6), Theorem 1a shows that $g_1 \in H^p$. Theorem 2c will follow once we show that $\check{G} = g_1$, as elements of S'.

From (2.7) and (2.8) it follows that $\langle g_1, \varphi \rangle = \langle \check{G}, \varphi \rangle$ for every $\varphi \in S$ which vanishes in a neighborhood of the origin. Hence $g_1 - \check{G}$ has support at the origin, and $\hat{g}_1 - G$ is a polynomial P.

Since $g_1 \in H^p$, we have

$$|g_1(\xi)| \leq C |\xi|^{n(\frac{1}{p} - 1)}$$

[TW, p. 105]. By (2.6), G also satisfies this bound, and hence so does P. Since $n(\frac{1}{p} - 1) \notin Z$, we must have $P \equiv 0$, and hence $g_1 = \check{G}$, as required.

The reader will note that Theorem 2c generalizes Theorem 2a, since it implies that if F is any function in $B_2^{n(\frac{1}{p} - \frac{1}{2}),p}$ then $G = F - P_N F$ has distributional Fourier transform in H^p, where $P_N F$ is the N'th

Taylor polynomial of F at $x = 0$.

We have been unable to find a substitute result for Theorem 2c when $n(\frac{1}{p} - 1)$ is an integer.

PROOF OF (2.2). Suppose that $F \in B_2^{\alpha,q}$. Write $f = \hat{F}$. By the definition (2.1) and Parseval's formula, (2.2) is equivalent to proving that the expression

$$(2.9) \qquad \|f\|_2 + \{\int_0^\infty t^{sq - \alpha q - 1} dt (\int_{\mathbb{R}^n} |x|^{2s} |f(x)|^2 e^{-t|x|} dx)^{q/2}\}^{1/q}$$

is equivalent to $\|f\|_{K_2^{\alpha,q}}$.

According to [F, Theorem 1], for $0 < k \leq \infty$, $0 < \mu < \infty$ and $h : (0,\infty) \to [0,\infty)$ we have

$$\int_0^\infty t^{\mu k - 1} dt (\int_0^\infty e^{-rt} h(r) dr)^k \approx \int_0^\infty t^{-\mu k - 1} dt (\int_t^{2t} h(r) dr)^k .$$

Apply this result with

$$h(r) = r^{n - 1 + 2s} \int_{|x| = r} |f(x)|^2 d\sigma(x) ,$$

$$k = q/2 , \quad \mu = 2s - 2\alpha , \quad d\sigma = \text{normalized surface measure} .$$

The second term in (2.9) is easily seen to be equivalent to

$$\{\int_0^\infty (\int_{t \leq |x| \leq 2t}^{2t} |f(x)|^2 |x|^{2s} dx)^{q/2} t^{q\alpha - qs - 1} dt\}^{1/q}$$

$$\approx \{\int_0^\infty (\int_{t \leq |x| \leq 2t}^{2t} |f(x)|^2 dx)^{q/2} t^{q\alpha - 1} dt\}^{1/q} ,$$

and this last expression is equivalent to

$$\left\{ \sum_{-\infty}^{\infty} \left(\int_{A_k} |f|^2 \right)^{q/2} 2^{k\alpha q} \right\}^{1/q} = \|f\|_{\dot{K}_2^{\alpha,q}} ,$$

as required.

3. MULTIPLIERS

If $f \in H^p$ then $\hat{f}$ is a continuous function satisfying $|\hat{f}(\xi)| \leq C \|f\|_{H^p} |\xi|^{n(\frac{1}{p} - 1)}$ [TW, p. 105]. It follows that if $m \in L^\infty(R^n)$ then $m\hat{f}$ defines an element of S', and thus the mapping $f \to (m\hat{f})^\vee$ is well defined from $H^p \to S'$. We say that m is a Fourier multiplier of H^p if this mapping takes H^p continuously into H^p.

The function m is said to satisfy a Hörmander condition of order s, where s is an integer, if

$$\int_{R<|\xi|<2R} |D^\beta m(\xi)|^2 \, d\xi \leq C R^{n-2|\beta|}, \qquad 0<R<\infty, \tag{3.1}$$

for all $|\beta| \leq s$.

Fix a function $\eta \in C_0^\infty(R^n)$ with

$$0 \leq \eta \leq 1, \quad \eta = 1 \text{ on } 1/2 \leq |\xi| \leq 2, \quad \operatorname{supp} \eta \subset \{1/4 \leq |\xi| \leq 4\}.$$

Define, for $\delta > 0$,

$$m_\delta(\xi) = m(\delta\xi)\eta(\xi).$$

In §1 we introduced certain spaces K and in §2 the Besov spaces B. It is easy to check, via Plancherel's theorem, that (3.1) is equivalent to

$$\sup_\delta \|\hat{m}_\delta\|_{K_2^{s,2}} < \infty, \tag{3.2}$$

which, by (2.2), is equivalent to

$$\sup_\delta \|m_\delta\|_{B_2^{s,2}} < \infty . \tag{3.3}$$

The following multiplier theorem is due to Calderón and Torchinsky [CT, p. 167]. A different proof was given by Taibleson-Weiss [TW, 9.45, 9.39]. Related results have been obtained by Peral and Torchinsky [PT].

CALDERON-TORCHINSKY THEOREM. *If* $\alpha > n(\frac{1}{p} - \frac{1}{2})$, $0 < p \leq 1$, *and if*

$$\sup_\delta \|\hat{m}_\delta\|_{K_2^{\alpha,2}} < \infty ,$$

or, equivalently,

$$\sup_\delta \|m_\delta\|_{B_2^{\alpha,2}} < \infty$$

then m *is a Fourier multiplier of* H^p, $0 < p \leq 1$.

It follows in particular that m is a Fourier multiplier of H^p if m satisfies a Hörmander condition of order $s > n(\frac{1}{p} - \frac{1}{2})$. The case $p = 1$ is due to Fefferman-Stein [FS, p. 150].

We are going to prove a sharp generalization of the Calderón-Torchinsky theorem which corresponds to the endpoint case for multiplier conditions in the same way that Theorem 1 corresponds to the endpoint case for embedding theorems. Note that, by (1.4) and (1.6), we have when $\alpha > n(\frac{1}{p} - \frac{1}{2})$ the continuous embeddings

$$K_2^{\alpha,2} \subset K_2^{n(\frac{1}{p} - \frac{1}{2}),p} \subset K_1^{n(\frac{1}{p} - 1),p} .$$

THEOREM 3a. *Suppose that* $0<p<1$ *and that*

$$M = \sup_{\delta} \|\hat{m}_{\delta}\|_{K_1^{n(\frac{1}{p}-1),p}} < \infty .$$

Then m *is a Fourier multiplier of* H^p , *and*

$$\|(m\,\hat{f})^{\vee}\|_{H^p} \leq CM\,\|f\|_{H^p} .$$

The theorem is sharp in the sense that $K_1^{n(\frac{1}{p}-1),p}$ cannot be replaced by any larger space of the K-type. The proof of the theorem is in §6 and the examples showing sharpness in §7.

From (2.2) and (1.4) - (1.6) it follows that $K_1^{n(\frac{1}{p}-1),p} \supset \mathcal{F}(B_2^{\alpha,q})$ if either $\alpha = n(\frac{1}{p}-\frac{1}{2})$ and $q \leq p$ or $\alpha > n(\frac{1}{p}-\frac{1}{2})$. Thus we have the following corollary, which furnishes a practical sufficient condition for multipliers.

COROLLARY 1. *Suppose that* $0<p<1$ *and that either* $\alpha = n(\frac{1}{p}-\frac{1}{2})$ *and* $q\leq p$ *or that* $\alpha > n(\frac{1}{p}-\frac{1}{2})$. *If*

$$\sup_{\delta} \|\hat{m}_{\delta}\|_{K_2^{\alpha,q}} < \infty$$

or, equivalently

$$\sup_{\delta} \|m_{\delta}\|_{B_2^{\alpha,q}} < \infty ,$$

then m *is a Fourier multiplier of* H^p .

The corollary is sharp in that $K_2^{n(\frac{1}{p}-\frac{1}{2}),p}$ cannot be replaced by any larger space $K_2^{\beta,q}$.

When $p=1$ Theorem 3a becomes false. The space $K_1^{0,1} = L^1$ has to be replaced by slightly smaller K-type spaces with appropriate weights $w(k)$. Suppose that $w: \{0,1,2,\dots\} \to [1,\infty)$ satisfies $1 \le w(k) \le w(k+1) < \infty$. Define the space $K(w)$ to be the set of all $f \in L^1(R^n)$ for which the norm

$$\|f\|_{K(w)} = \int_{|x|<1} |f| + \sum_{k=0}^{\infty} \left(\int_{A_k} |f|\right) w(k)$$

is finite. A critical role is played by the condition

$$\sum_{k=0}^{\infty} w(k)^{-2} < \infty . \tag{3.4}$$

THEOREM 3b. *If* w *satisfies* (3.4) *and if*

$$M = \sup_{\delta} \|\hat{m}_\delta\|_{K(w)} < \infty \tag{3.5}$$

then m *is a Fourier multiplier of* H^1, *and*

$$\|(m\hat{f})^{\vee}\|_{H^1} \le C M \|f\|_{H^1}$$

where $C = C(n)\left(\sum_{k=1}^{\infty} w(k)^{-2}\right)^{1/2}$.

In particular, we may take $w(k) = 2^{\varepsilon k}$, $\varepsilon > 0$. Then $K(w) = K_1^{\varepsilon,1}$. Since $K_2^{\alpha,2} \subset K_1^{\alpha-\frac{n}{2},2} \subset K_1^{\varepsilon,1}$ when $0 < \varepsilon < \alpha - \frac{n}{2}$, the Calderón-Torchinsky multiplier theorem for $p=1$ follows from Theorem 3b.

Theorem 3b is sharp, at least within the context of reasonable $w(k)$'s. We show in §7 that if $w(k)\uparrow$, $w(2k) \le C\,w(k)$, and if $\sum_{k=0}^{\infty} w(k)^{-2} = \infty$ then there exists m satisfying (3.5) and $f \in H^1$ for which $(m\hat{f})^{\vee} \notin H^1$. These examples will also show that in the Calderón-Torchinsky theorem $K_2^{\alpha,2}$ cannot be replaced by $K_2^{n/2,q}$, no matter how small q is.

If no growth restriction is placed on $w(k)$ the situation becomes rather chaotic. There exists a positive nondecreasing sequence $w(k)$ such that $\sum w(k)^{-2} = \infty$, but the only $m \in L^\infty(\mathbb{R}^n)$ satisfying (3.5) is $m \equiv 0$. We will not give this construction, but only point out that the main fact used is that if $g \in L^1$ and $\operatorname{supp} g \subset \{|x| < 1\}$, then $\hat{g}(x+iy)$ is an entire function in $\mathbb{C}^n$ which is bounded on sets $\{z \in \mathbb{C}^n : |\operatorname{Im} z| \leq M\}$.

Suppose that w satisfies (3.4) and m satisfies (3.5). Then $m \in L^\infty$, since $K(w) \subset L^1$. Thus m is a Fourier multiplier on L^2. For $1<p<2$ L^p is an interpolation space between H^1 and L^2 ([FS, p. 156]). Thus m Fourier multiplies L^p for $1<p\leq 2$, and by duality this is also true for $2\leq p<\infty$. Thus we arrive at the following refinement of Hörmander's multiplier theorem.

COROLLARY 2. *If m and w satisfy the hypotheses of Theorem 3b, then m is a Fourier multiplier on L^p, $1<p<\infty$.*

Hörmander's theorem, as stated in [S, p. 96, Cor] is implied by the special cases $w(k) = 2^{\epsilon k}$, $\epsilon > 0$ sufficiently small.

As an application of Theorem 3a, we shall use it to prove a recent theorem of Miyachi [M] about "strongly singular" multipliers. Suppose that $0<p<1$, $a>0$, $b>0$ and that s is an integer larger than $\frac{b}{a}$. Assume that $m(\xi)$ satisfies

$$(3.6) \qquad |D^\beta m(\xi)| \leq C\, |\xi|^{-b}\, |\xi|^{(a-1)|\beta|}, \quad 0 \leq |\beta| \leq s,$$

$$m(\xi) = 0 \quad \text{for} \quad |\xi| < 1.$$

The prototypical example is the function $m_{a,b}$ defined by

$$m_{a,b}(\xi) = \psi(\xi)\, |\xi|^{-b}\, e^{i|\xi|^a}$$

where $\psi \in C^\infty$, $\psi = 0$ on $|\xi|<1$, $\psi = 1$ on $|\xi| \geq 2$.

Miyachi proved that (3.6) implies that m is a Fourier multiplier of H^p provided

$$n(\frac{1}{p} - \frac{1}{2}) \leq \frac{b}{a} . \tag{3.7}$$

For $m_{a,b}$ with a and b satisfying certain restrictions this had been proved earlier by Sjölin [Sj]. The case $1 \leq p < 2$, $0 < a < 1$ had been settled by Fefferman-Stein [FS, p. 160]. Miyachi shows by an interpolation argument that the results for $1 \leq p < 2$ are consequences of those for $0 < p < 1$. His Theorem 3 shows that $m_{a,b}$ does not Fourier multiply H^p when $n(\frac{1}{p} - \frac{1}{2}) > \frac{b}{a}$, $0 < p < 1$.

Suppose that $0 < p < 1$ and that (3.6) holds. Set $\alpha = \frac{b}{a}$. By Corollary 1 and (2.2), Miyachi's theorem will follow from the estimates

$$\sup_{\delta} \|m_\delta\|_{B_2^{\alpha,p}} < \infty . \tag{3.8}$$

Fix δ, and write $g(\xi) = m_\delta(\xi) = m(\delta\xi)\eta(\xi)$. Then $g \equiv 0$ if $\delta \leq \frac{1}{4}$ so we assume from now on that $\delta \geq \frac{1}{4}$. Hypothesis (3.6) implies that

$$\|D^\beta g\|_\infty \leq C\, \delta^{a|\beta| - b}, \quad 0 \leq |\beta| \leq s , \tag{3.9}$$

where C is independent of δ.

Assume temporarily that $\alpha \notin \mathbb{Z}$ and write $\alpha = \nu + \sigma$, where ν is a non-negative integer and $0 < \sigma < 1$. Then $\nu < s$. By [S, p. 153], (3.8) is equivalent to the estimates

$$\|D^\beta g\|_{L^2} \leq C , \quad 0 \leq |\beta| \leq \nu , \tag{3.10}$$

and

$$\int_{R^n} \|\Delta_t h\|_2^p \, |t|^{-p\sigma - n} \, dt \le C \, , \tag{3.11}$$

where h denotes a partial derivative of g of order ν and $(\Delta_t h)(x) = h(x+t) - h(x)$.

Since g has support in $\frac{1}{2} \le |x| \le 2$, (3.10) follows from (3.9). As for (3.11),

$$\|\Delta_t h\|_\infty \le C \, |t| \max_{|\beta| = \nu + 1} \|D^\beta g\|_\infty \le C \, |t| \delta^{a(\nu+1) - b} \tag{3.12}$$

and also

$$\|\Delta_t h\|_\infty \le C \, \|h\|_\infty \le C \, \delta^{a\nu - b} \, . \tag{3.13}$$

For fixed t, $\Delta_t h(x) = 0$ except on a set of measure $\le C$. Hence (3.12) and (3.13) are true with L^2 norms in place of L^∞ norms. Using (3.12) for $|t| \le \delta^{-a}$ and (3.13) for $|t| \ge \delta^{-a}$, it follows that the integral in (3.11) is dominated by C times

$$\int_{|t| \le \delta^{-a}} |t|^{p(1-\sigma) - n} \delta^{p[a(\nu+1) - b]} \, dt + \int_{|t| \ge \delta^{-a}} |t|^{-p\sigma - n} \delta^{(a\nu - b)p} \, dt$$

$$= C \, \delta^{p(a\sigma + a\nu - b)} = C \, ,$$

since $\alpha = \frac{b}{a} = \nu + \sigma$. This proves Miyachi's theorem when $\alpha \notin Z$.

If $\alpha \in Z$ write $\alpha = \nu + 1$, $\nu \ge 0$. Then (3.8) is equivalent to (3.10) and

$$\int_{R^n} \|\Delta_t^2 h\|_2^p \, |t|^{-p - n} \, dt \le C \tag{3.14}$$

where $(\Delta_t^2 h)(\xi) = h(\xi + t) + h(\xi - t) - 2h(\xi)$. The estimate (3.14) is accomplished as above, the only change being that (3.12) is replaced by

$$\|\Delta_t^2 h\|_\infty \leq C\,|t|^2 \max_{|\beta| = \nu + 2} \|D^\beta g\|_\infty .$$

The Calderón-Torchinsky theorem implies that an m satisfying (3.6) will Fourier multiply H^p for $n(\frac{1}{p} - \frac{1}{2}) < \frac{b}{a}$, but to get boundedness at the critical index p_0, $n(\frac{1}{p_0} - \frac{1}{2}) = \frac{b}{a}$, it seems essential to have a multiplier theorem for the endpoint case $K_2^{n(\frac{1}{p} - \frac{1}{2}),p}$.

4. PROOF OF THEOREM 1

Suppose that $0<p\leq 1$ and that $f\in \dot{K}_1^{n(\frac{1}{p}-1),p}$ if $0<p<1$ and $n(\frac{1}{p}-1)\notin Z$, $f\in Y(p)$ if $p<1$ and $N=n(\frac{1}{p}-1)$, $f\in Y^*$ if $p=1$.

Take $\psi\in C_0^\infty(R^n)$ with $\operatorname{supp}\psi\subset\{|x|<1\}$; $\int\psi=1$. Set $\psi_{x,t}(y)=t^{-n}\psi(\frac{x-y}{t})$. Define

$$g(x,t)=(f*\psi_t)(x)=\langle f,\psi_{x,t}\rangle.$$

By the definitions (1.9), (1.10) and those of $Y(p)$ and Y^* we have

$$g(x,t)=\int f(y)(\psi_{x,t}(y)-(P_N\psi_{x,t})(y))\,dy\,. \tag{4.1}$$

Our aim is to show that $\sup_{t>0}|g(x,t)|\in L^p$, with appropriate bounds.

We claim that

$$\sup_{t>0}|g(x,t)|\leq C\,(J_1(x)+J_2(x)+J_3(x)) \tag{4.2}$$

where

$$J_1(x)=\sup_{t<\frac{1}{2}|x|}t^{-n}\int_{|y-x|\leq t}|f(y)|\,dy\,,$$

$$J_2(x)=\int_{|y|>2|x|}|f(y)|\left(\frac{|y|}{|x|}\right)^N|x|^{-n}\,dy\,,$$

$$J_3(x)=\int_{|y|<2|x|}|f(y)|\left(\frac{|y|}{|x|}\right)^{N+1}|x|^{-n}\,dy\,.$$

For $N = n(\frac{1}{p} - 1)$ we also need two other estimates.

$$\sup_{t>0} |g(x,t)| \leq C\,(J_1(x) + J_4(x) + J_5(x))\ , \qquad N = n(\tfrac{1}{p} - 1)\ , \qquad p<1\ , \tag{4.3}$$

where

$$J_4(x) = \int_{|y|>2|x|} |f(y)| \left(\frac{|y|}{|x|}\right)^{N-1} |x|^{-n}\, dx\ ,$$

$$J_5(x) = \int_{|y|<2|x|} |f(y)| \left(\frac{|y|}{|x|}\right)^{N} |x|^{-n}\, dx\ ,$$

$$\sup_{t>0} |g(x,t)| \leq C\,(J_6(x) + \|f\|_{L^1})\ , \qquad p=1\ , \tag{4.4}$$

where

$$J_6(x) = \sup_{t<1} t^{-n} \int_{|y-x|<t} |f(y)|\, dy\ .$$

The proofs of (4.2) - (4.4) are at the end of the section.

Assume now that $n(\frac{1}{p} - 1) > N$. We shall prove Theorem 1a by showing that $J_i \in L^p$, $i = 1,2,3$, with appropriate bounds. Consider first J_2.

Suppose that $x \in A_k = \{x : 2^k \leq |x| \leq 2^{k+1}\}$, $-\infty<k<\infty$. Then, writing

$$m_j = \int_{A_j} |f|\, dx\ , \tag{4.5}$$

we have

$$J_2(x) \leq C \sum_{j=k+1}^{\infty} m_j\, 2^{jN}\, 2^{-k(N+n)}\ ,$$

$$\int_{A_k} J_2(x)^p\, dx \leq C \sum_{j=k+1}^{\infty} m_j^p\, 2^{jNp}\, 2^{-kp(N+n)}\, 2^{kn}\ ,$$

$$\sum_{k=-\infty}^{\infty} \int_{A_k} J_2(x)^p \, dx \leq C \sum_{j=-\infty}^{\infty} \sum_{k=-\infty}^{j-1} m_j^p \, 2^{jNp} \, 2^{k[n-p(N+n)]} .$$

Since $n(\frac{1}{p} - 1) > N$, we deduce

$$(4.6) \quad \int_{R^n} J_2(x)^p \, dx \leq C \sum_{j=-\infty}^{\infty} m_j^p \, 2^{jNp} \, 2^{j[n-p(N+n)]} = C \, \|f\|^p_{\dot{K}^{n(\frac{1}{p}-1),p}} .$$

A completely analogous computation yields

$$(4.7) \qquad \int_{R^n} J_3(x)^p \, dx \leq C \, \|f\|^p_{\dot{K}_1^{n(\frac{1}{p}-1),p}} , \qquad \text{when } n(\tfrac{1}{p} - 1) > N .$$

To estimate $\int J_1^p \, dx$ we need some inequalities for the Hardy-Littlewood maximal function. For $F \in L^1(R^n)$ write

$$(MF)(x) = \sup_{x \in B} \frac{1}{|B|} \int_B |F| \, dx , \qquad x \in R^n ,$$

where B denotes a ball.

LEMMA. <u>Suppose</u> $F \in L^1(R^n)$ <u>and that</u> F <u>is supported in a ball</u> B. <u>Then</u>

(a) $\int_B (MF)^p \, dx \leq C_p \, (\int_B |F|)^p \, |B|^{1-p}$, $\quad 0 < p < 1$,

(b) $\int_B MF \, dx \leq C \int_B |F| \, [1 + \log^+ \left(\frac{|F| \, |B|}{\int_B |F|}\right)] \, dx$.

<u>Proof</u>. A homogeneity argument shows that it suffices to consider the case when $\int_B |F| = 1$ and $|B| = 1$. Then the inequality in (a) follows from the weak 1-1 estimate

$$|\{x \in R^n : MF(x) \geq \alpha\}| \leq C\alpha^{-1} \int_B |F| \, dx$$

and (b) follows from the sharper estimate

$$|\{x \in R^n : (MF)(x) \geq \alpha\}| \leq C\alpha^{-1} \int_{|F| > \frac{1}{2}|\alpha|} |F| \, dx \, .$$

See [S, p. 23].

Fix $k \in \mathbb{Z}$ and define $F(x) = f(x)$ if $x \in \bar{A}_k = \{x : 2^{k-1} < |x| < 2^{k+2}\}$, $F(x) = 0$ otherwise. Then $J_1(x) \leq C \, MF(x)$ for $x \in A_k$. For $0 < p < 1$, the lemma gives

$$\int_{A_k} J_1^p \leq C \int_{|x| < 2^{k+2}} (MF)^p \, dx \leq C(\int_{\bar{A}_k} |F|)^p \, 2^{nk(1-p)}$$

$$\leq C 2^{nk(1-p)} (m_{k-1}^p + m_k^p + m_{k+1}^p) \, .$$

Hence

$$\int_{\mathbb{R}^n} J_1^p \, dx \leq C \, \|f\|^p_{\dot{K}_1^{n(\frac{1}{p}-1),p}} \, , \qquad 0 < p < 1 \, . \tag{4.8}$$

Theorem 1a now follows from (4.6), (4.7) and (4.8).

Assume next that $0 < p < 1$, and that $n(\frac{1}{p} - 1) = N$. We use (4.2) for $|x| < 1$ and (4.3) for $|x| < 1$.

Thus

$$\int_{\mathbb{R}^n} (\sup_{t>0} |g(x,t)|)^p \, dx \leq C \, (\int_{\mathbb{R}^n} J_1^p \, dx + \int_{|x|>1} J_2^p \, dx + \int_{|x|>1} J_3^p \, dx \tag{4.9}$$

$$+ \int_{|x|>1} J_4^p \, dx + \int_{|x|<1} J_5^p \, dx) \, .$$

Now (4.8) is still valid, so the first term on the right is suitably bounded. Next,

$$(4.10)\quad \int_{|x|>1} J_2^p\,dx \le C \sum_{k=0}^{\infty} \int_{A_k} J_2^p\,dx \le C \sum_{k=0}^{\infty} \sum_{j=k+1}^{\infty} m_j^p\, 2^{jNp}\, 2^{k[n-p(N+n)]}$$

$$= C \sum_{j=1}^{\infty} \sum_{k=0}^{j-1} m_j^p\, 2^{jNp} = C \sum_{j=1}^{\infty} j\, m_j^p\, 2^{jNp} \le C\, \|f\|_{Y(p)}^p \ .$$

Similarly,

$$(4.11)\quad \int_{|x|>1} J_3^p\,dx \le C \sum_{k=0}^{\infty} \sum_{j=-\infty}^{k+2} m_j^p\, 2^{j(N+1)p}\, 2^{k[n-(n+N+1)p]}$$

$$= C\Big(\sum_{j=-\infty}^{1} \sum_{k=0}^{\infty} + \sum_{j=2}^{\infty} \sum_{k=j-2}^{\infty} m_j^p\, 2^{j(N+1)p}\, 2^{-kp}\Big)$$

$$\le C\Big(\sum_{j=-\infty}^{1} m_j^p\, 2^{j(N+1)p} + \sum_{j=2}^{\infty} m_j^p\, 2^{jNp}\Big)$$

$$\le C\, \|f\|_{Y(p)}^p \ .$$

Also, when $n(\frac{1}{p} - 1) = N$,

$$(4.12)\quad \int_{|x|<1} J_4^p\,dx \le C\, \|f\|_{Y(p)}^p \ ,$$

$$(4.13)\quad \int_{|x|<1} J_5^p\,dx \le C \sum_{j=-\infty}^{0} (|j|+1)\, m_j^p\, 2^{jNp} \le C\, \|f\|_{Y(p)}^p \ .$$

Theorem 4b follows from (4.8) - (4.13) .

Now assume $p = 1$. In proving Theorem 1c we may assume that

$$\|f\|_{L^1} = 1 \ .$$

By (4.2) and (4.4) we have

(4.14) $$\|f\|_{H^1} \le C \int_{|x|>1} (J_1 + J_2 + J_3)\,dx + C \int_{|x|<1} (J_6 + 1)\,dx .$$

Computations like the ones above give

(4.15) $$\int_{|x|>1} J_2\,dx \le C \sum_{j=1}^{\infty} j\, m_j \le C \int_{|x|>1} |f(x)|(1+\log|x|)\,dx \le C\|f\|_{Y^*} ,$$

(4.16) $$\int_{|x|>1} J_3\,dx \le C\,\Big(\sum_{j=-\infty}^{0} m_j\, 2^j + \sum_{j=1}^{\infty} m_j\Big) \le C \int_{R^n} |f| \le C\,\|f\|_{Y^*} .$$

Next, define again $F(x) = f(x)$ for $x \in \bar{A}_k = \{2^{k-1} \le |x| \le 2^{k+2}\}$ $F(x) = 0$ for $x \notin \bar{A}_k$, where $k \ge 0$ is fixed. Then $J_1 \le C\,MF$ on A_k. By part (b) of the lemma, we have, recalling $\|f\|_{L^1} = 1$,

$$\int_{A_k} J_1\,dx \le C \int_{|x|<2^{k+2}} MF\,dx \le C \int_{\bar{A}_k} |F| \left[1 + \log^+\left(\frac{|F|\,C2^{nk}}{\int_{\bar{A}_k} |F|}\right)\right] dx$$

$$\le C \int_{\bar{A}_k} |f| \left[1 + \log^+|f| + k + \log \frac{1}{\int_{\bar{A}_k} |f|}\right] dx .$$

Let $S_1 = \{k \ge 0 : \int_{\bar{A}_k} |f| \ge e^{-k}\}$, $S_2 = \{k \ge 0 : \int_{\bar{A}_k} |f| < e^{-k}\}$. Then, for $k \in S_1$,

$$\int_{A_k} J_1\,dx \le C \int_{\bar{A}_k} |f|(1 + \log^+|f| + k)\,dx .$$

Since $x \log \frac{1}{x} \le x^{1/2}$ for $0 < x < 1$, we have

$$\Big(\int_{\bar{A}_k} |f|\, dx\Big)\log \frac{1}{\int_{\bar{A}_k} |f| dx} \le e^{-k/2}, \quad \text{if } k \in S_2 .$$

So, for $k \in S_2$,

$$\int_{A_k} J_1\, dx \le \int_{\bar{A}_k} |f|[1 + \log^+|f| + \log|x|]\, dx + e^{-k/2} .$$

Adding up, and using $\|f\|_{L^1} = 1$, we find

$$(4.17) \quad \int_{|x|>1} J_1\, dx \le C \int_{R^n} |f|(1 + \log^+|f| + \log^+|x|)\, dx + 3 \le C\, \|f\|_{Y^*} .$$

Finally, to estimate J_6 we let $F(x) = f(x)$ if $|x| < 2$, $F(x) = 0$ if $|x| \ge 2$. Then $J_6(x) \le C\, MF(x)$ for $|x| < 1$ and, by part (b) of the Lemma,

$$\int_{|x|<1} J_6\, dx \le C \int_{|x|<1} (MF)\, dx \le C \int_{|x|<2} |F(x)|\Big[1 + \log^+\Big(\frac{|F(x)|C}{\int_{|x|<2} |F|}\Big)\Big]\, dx$$

$$\le C \int_{|x|<2} |f(x)|\Big[1 + \log^+|f(x)| + \log\Big(\frac{1}{\int_{|x|<2} |f|}\Big)\Big]\, dx$$

$$\le C\, \|f\|_{Y^*} + C\Big(\int_{|x|<2} |f|\Big) \log\Big(\frac{1}{\int_{|x|<2} |f|}\Big) \le C\, \|f\|_{Y^*} ,$$

where the last estimate uses $1 = \|f\|_{L^1} \le \|f\|_{Y^*}$.

Hence

$$(4.18) \quad \int_{|x|<1} (J_6(x) + 1)\, dx \le C\, \|f\|_{Y^*} .$$

Theorem 1c follows from (4.14) - (4.18) .

It remains to prove (4.2) - (4.4) . Write

$$R(x,y,t) = \psi_{x,t}(y) - (P_N \psi_{x,t})(y) .$$

Then

$$g(x,t) = \int f(y)\, R(x,y,t)\, dy .$$

We need the following estimates for R .

(4.19) If $t < \frac{1}{2}|x|$ then $|R| \leq Ct^{-n}$.

(4.20) If $t > \frac{1}{2}|x|$ and $|y| < 4t$, then $|R| \leq Ct^{-n}(\frac{|y|}{t})^{N+1}$.

(4.21) If $t > \frac{1}{2}|x|$ and $|y| \geq 4t$ then $|R| \leq Ct^{-n}(\frac{|y|}{t})^{N}$.

Recall that $\psi_{x,t}(y) = t^{-n}\psi(\frac{x-y}{t})$. Suppose for simplicity that $y = (|y|, 0, \ldots, 0)$. Then

$$(P_N \psi_{x,t})(y) = \sum_{j=0}^{N} \frac{1}{j!} \frac{\partial^j \psi_{x,t}(0)}{\partial y_1^j} |y|^j = t^{-n} \sum_{j=0}^{N} \frac{(-1)^j t^{-j}}{j!} \frac{\partial^j \psi}{\partial y_1^j}(\tfrac{x}{t})\, |y|^j .$$

PROOF OF (4.19). Since $\operatorname{supp} \psi \subset (|x| < 1)$ we have $P_N \psi_{x,t}(y) = 0$. Hence $|R| = |\psi_{x,t}| \leq Ct^{-n}$.

PROOF OF (4.20). By Taylor's theorem,

$$|R(x,y,t)| \leq C|y|^{N+1} \sup_{0<z<|y|} \left|\frac{\partial^{N+1}}{\partial y_1^{N+1}} \psi_{x,t}(z)\right| \leq C|y|^{N+1}\, t^{-(N+1)}\, t^{-n} ,$$

since $\psi \in C_0^\infty$.

PROOF OF (4.21). Since supp $\psi \subset \{|x| < 1\}$ we have $\psi_{x,t}(y) = 0$. Hence

$$|R(x,y,t)| = |(P_N \psi_{x,t})(y)| \leq C(\frac{|y|}{t})^N t^{-n} .$$

PROOF OF (4.2). Suppose that $t > \frac{1}{2}|x|$. Using (4.20) and (4.21) we obtain

$$|g(x,t)| \leq C \int_{|y|<4t} |f(y)| \, t^{-n}(\frac{|y|}{t})^{N+1} dy + C \int_{|y|>4t} |f(y)| \, t^{-n}(\frac{|y|}{t})^N dy$$

$$\leq C \int_{|y|<2|x|} |f(y)| \left(\frac{|y|}{|x|}\right)^{N+1} |x|^{-n} dy + C \int_{2|x|<|y|<4t} |f(y)| \, |x|^{-n} dy$$

$$+ C \int_{|y|>2|x|} |f(y)| \, |x|^{-n} \left(\frac{|y|}{|x|}\right)^N dy$$

$$\leq C \, (J_2(x) + J_3(x)) , \quad t > \frac{1}{2}|x| ,$$

since the second integral is dominated by $C\, J_2(x)$.

If $t < \frac{1}{2}|x|$ then (4.19) shows that

$$|g(x,t)| \leq C \, J_1(x) ,$$

which, with the previous estimate, proves (4.2).

PROOF OF (4.3). Suppose that $0 < p < 1$, $N = n(\frac{1}{p} - 1)$. If $f \in Y(p)$ then by definition (1.10)

$$g(x,t) = \int f(y)(\psi_{x,t}(y) - (P_{N-1} \psi_{x,t})(y)) \, dy ,$$

and (4.3) follows by replacing N by $N-1$ in the argument above.

PROOF OF (4.4). If $f \in Y^* \subset L^1$ then

$$g(x,t) = \int f(y)\, \psi_{x,t}(y)\, dy ,$$

and we obtain

$$\sup_{t>0} |g(x,t)| \le \sup_{t\le 1} |g(x,t)| + \sup_{t\ge 1} |g(x,t)|$$

$$\le \sup_{t<1} Ct^{-n} \int_{|y-x|<t} |f(y)|\, dy + C \int_{\mathbb{R}^n} |f(y)|\, dy$$

$$= C(J_6(x) + \|f\|_{L^1}) .$$

5. BEST POSSIBLE NATURE OF THEOREMS 1B AND 1C

In §1 we gave simple examples illustrating the sharpness of Theorem 1a. Here we consider the case when $n(\frac{1}{p} - 1) \in Z$. Throughout this section $p \in (0,1]$ and $N \in Z$ will be connected by

$$N = n(\frac{1}{p} - 1), \qquad N = 0,1,2,\ldots .$$

In theorems 4a and 4b $\varepsilon(k)$, $k \geq 0$ stands for an arbitrarily prescribed sequence with

$$\varepsilon(k) \geq \varepsilon(k+1), \quad \lim_{k\to\infty} \varepsilon(k) = 0 .$$

THEOREM 4a. *For each* $N \geq 0$ *there is a function* $f_1 \in L^1$ *satisfying*

(5.1) $\quad f_1(x) = 0$ for $|x| < 1$, $\int |f_1(x)||x|^N dx < \infty$, *and* (1.1),

such that

(5.2) $\quad f_1 \in K_a^{n(\frac{1}{p} - \frac{1}{a}),q}$, *for every* $1 \leq a \leq \infty$ *and* $q > 0$,

(5.3) $\quad \sum_{k=0}^{\infty} (\int_{A_k} |f_1|^p \, 2^{nk(1-p)}) |k| \varepsilon(k) < \infty$,

but $f_1 \notin H^p$.

THEOREM 4b. *For each* $N \geq 0$ *there is a function* f_2 *satisfying*

$$f_2(x) = 0 \quad \text{for} \quad |x| > 1 , \quad \int |f_2(x)|\,|x|^N \, dx < \infty , \quad \text{and} \quad (1.13) ,$$

such that

$$\sum_{k=-\infty}^{-1} \left(\int_{A_k} |f_2|^p\right) 2^{nk(1-p)} \, |k| \, \varepsilon(-k) < \infty ,$$

but the distribution defined by f_2 *via* (1.10) *does not belong to* H^p .

THEOREM 4c. *Given a non-decreasing function* $L(x)$, $x > 0$, *satisfying* $L(0) \geq 1$ *and*

$$L(x) = o(\log x) , \quad x \to \infty ,$$

there exists $f_3 \in L^1(\mathbb{R}^n)$ *satisfying*

$$f_3(x) = 0 \quad \text{for} \quad |x| > 1 , \quad \|f_3\|_{L^1} = 1 , \quad \text{and} \quad \int f_3 = 0 ,$$

such that

$$\int |f_3| \, L(|f_3|) \, dx < \infty ,$$

but $f_3 \notin H^1$.

PROOF OF THEOREM 4a. Write $e_1(N) = (1,0,\ldots,0) \in \mathbb{R}^N$ and choose $f_0 \in L^\infty$ with the following properties:

$$\operatorname{supp} f_0 \subset A_0 , \quad \int f_0(x) x_1^N \, dx = \delta > 0 ,$$

$$\int_{|x|=1} f_0(rx) x^\beta \, d\sigma(x) = 0 , \quad 0 \leq |\beta| \leq N , \quad \beta \neq e_1(N) , \quad 0 < r < \infty ,$$

where $d\sigma$ denotes surface measure. For $N=0$ the vanishing moment condition is vacuous.

Next, let S be an infinite subsequence of $\{4^{\ell} : \ell \in Z,\ \ell \geq 2\}$ for which $\sum_{k \in S} \epsilon(k) < \infty$.

Define

$$f_1(x) = \begin{cases} \alpha_k f_0(2^{-k}x), & x \in A_k, \quad k \in S, \\ -\alpha_k f_0(2^{-2k}x)\, 2^{-k(n+N)}, & x \in A_{2k}, \quad k \in S, \\ 0, & \text{otherwise} \end{cases}$$

where

$$\alpha_k = (k^{-1}\, 2^{-kn})^{1/p} \qquad (5.4)$$

The lacunarity of S implies (5.2), and the reader may easily verify (5.1) and (5.3). We shall prove $f_1 \notin H^p$.

Choose $\psi \in C^{\infty}$ with the properties

$$\operatorname{supp} \psi \subset (|x| < 6), \quad \psi(x) = x_1^N \text{ if } |x| < 5, \quad \int \psi = 1,$$

and define

$$g(x,t) = \int f_1(y)\psi\left(\frac{x-y}{t}\right) t^{-n}\, dy .$$

Suppose that $k \in S$ and $k \leq j \leq 2k-4$. Then for $x \in A_j$ we have

$$g(x,|x|) = \int_{A_k} f_1(y)(-1)^N y_1^N |x|^{-N-n} dy$$

$$= (-1)^N \delta\, 2^{k(N+n)} |x|^{-(N+n)} \alpha_k ,$$

$$|g(x,|x|)| \geq C\, 2^{k(N+n)}\, 2^{-j(N+n)} \alpha_k ,$$

$$\int_{A_j} |g(x,|x|)|^p dx \geq C\, 2^{kp(N+n)}\, 2^{-jp(N+n)} \alpha_k^p\, 2^{nj}$$

$$= C\, \alpha_k^p\, 2^{kn}$$

$$= C\, k^{-1} ,$$

where $C > 0$. The last equation follows from (5.4) and the next-to-last one from $p(N+n) = n$.

Hence, for $k \in S$,

$$\int_{2^k < |x| < 2^{2k}} |g(x,|x|)|^p dx \geq C ,$$

so that $\int |g(x,|x|)|^p dx = \infty$ and $f_1 \notin H^p$.

PROOF OF THEOREM 4b. Let f_0 and S be as in the previous proof, and define

$$f_2(x) = \begin{cases} \alpha_k f_0(2^{-k}x) , & x \in A_k , \quad (-k) \in S , \\ -\alpha_k f_0(2^{-2k}x)\, 2^{-k(n+N)} , & x \in A_{2k} , \quad (-k) \in S , \\ 0 , & \text{otherwise} . \end{cases}$$

The proof that f_2 has the desired properties proceeds as above, the only change being in the definition of $g(x,t)$. This time

$$g(x,t) = \int f_2(y)[\psi_{x,t}(y) - (P_{N-1}\psi_{x,t})(y)]\,dy ,$$

where ψ is as before, P_{N-1} denotes $(N-1)$st Taylor polynomial at $y=0$, and $\psi_{x,t}(y) = t^{-n}\psi(\frac{x-y}{t})$.

PROOF OF THEOREM 4c. Choose numbers $x_j > 2^{2n}$, $j = 1,2,3,\ldots,$ such that

$$x_j\, L(x_j) < 2^{-j}\, x_j \log x_j \tag{5.5}$$

and

$$2^{n+1}\, x_j \log x_j < x_{j+1} \log x_{j+1} . \tag{5.6}$$

Define integers $k(j)$ by

$$2^{nk(j)}\, x_j \log x_j \in [\tfrac{1}{2}, 1) . \tag{5.7}$$

Then $k(j+1) \leq k(j) - 2$ and $k(1) \leq -3$. Define

$$g(x) = \begin{cases} x_j , & \text{if } x \in A_{k(j)} , \quad j = 1,2,\ldots, \\ 0 , & \text{otherwise} \end{cases}$$

It follows from (5.5) and (5.7) that

$$\int g\, L(g) < \infty .$$

Notice that g is supported in $|x| \leq \frac{1}{4}$. Let $g_1 = g + b$, where $b \in L^\infty$, $\operatorname{supp} b \subset (\frac{1}{4} < |x| < \frac{1}{2})$, and $\int b = \int -g$. If $\|g_1\| \geq 1$ define

$f_3 = \dfrac{g_1}{\|g_1\|_1}$. If $\|g_1\| < 1$ let $f_3 = g_1 + b_1$, where $b_1 \in L^\infty$ is chosen so that $\operatorname{supp} b_1 \subset (\frac{1}{2} < |x| < 1)$, $\int b_1 = 0$, and $\|g_1 + b_1\|_1 = 1$. Then $\int f_3 = 0$, $\|f_3\|_{L^1} = 1$, and

$$\int |f_3|\ L(f_3|) < \infty .$$

Take $\psi \in C^\infty$ with $\psi \geq 0$, $\psi = 1$ on $|x| < \frac{1}{2}$, $\operatorname{supp} \psi \subset (|x| < 1)$, $\int \psi\, dx = 1$. Define

$$u(x,t) = \int f_3(y)\, \psi(\frac{x-y}{t})\, t^{-n}\, dy ,$$

$$N(x) = \sup_{|y-x| \leq t} |u(y,t)| .$$

Then

$$N(x) \geq u(0,|x|) \geq C\, |x|^{-n} \int_{|y| < \frac{1}{2}|x|} g\, dy - C .$$

Write $m_\ell = \int_{A_\ell} g$. Then

$$\int_{A_k} N\, dx \geq C \sum_{\ell \leq k-1} m_\ell - C2^{nk} ,$$

$$\int_{|x|<1} N\, dx \geq C \sum_{\ell=-\infty}^{-2} (|\ell| - 1) m_\ell - C$$

$$\geq C \sum_{j=1}^{\infty} |k(j)|\, x_j\, 2^{nk(j)} - C .$$

By (5.5) and (5.7) we have

$$x_j\, L(x_j)\, 2^{nk(j)} \leq 2^{-j}\, x_j(\log x_j)\, 2^{nk(j)} \leq 2^{-j} < 1 .$$

Since $L(x_j) \geq 1$, it follows that

$$\log x_j < C\, n|k(j)| .$$

Hence

$$\int_{|x|<1} N\, dx \geq C \sum_{j=1}^{\infty} x_j(\log x_j)\, 2^{nk(j)} = \infty ,$$

where we have used (5.7). Thus $N \notin L^1$, and $f_3 \notin H^1$, by Theorem 4 in [FS, p. 151].

6. PROOF OF THEOREM 3

Our proof makes use of the atomic decomposition of H^p. The function a is said to be a p-atom if

$$\text{supp } a \subset B, \quad \text{for some ball } B \subset R^n,$$

(6.1) $$|a(x)| \leq |B|^{-1/p}$$

$$\int x^\beta a(x)\,dx = 0, \quad \text{for } 0 \leq |\beta| \leq N.$$

Recall that $N = [n(\frac{1}{p} - 1)]$.

These are (p,∞,N) atoms in the notation of [TW].

Coifman [C], when $n = 1$, and Latter [L], for general n, proved that $f \in S'$ belongs to H^p if and only if f admits a representation $f = \sum_1^\infty \lambda_j a_j$, with convergence in S', where the a_j are p-atoms and the λ_j are constants with $\sum_1^\infty |\lambda_j|^p < \infty$. Moreover, $\|f\|_{H^p}^p$ is comparable to $\inf \Sigma|\lambda_j|^p$, the inf being taken over all possible such representations.

To prove that the operator $f \to (m\hat{f})^\vee$, $m \in L^\infty$, takes H^p continuously into H^p it suffices to show that $\|(m\hat{a})^\vee\|_{H^p} \leq C$ for all p-atoms a. We may assume that the ball B of (6.1) has center at the origin. Furthermore, since the number $\sup_\delta \|\hat{m}_\delta\|_X$ is the same for all dilations of m, to obtain Theorems 3a and 3b it will be enough to consider the case when B is $|x| < 1$.

Thus, our problem is reduced to proving that

(6.2) $$\|(m\hat{a})^\vee\|_{H^p} \leq C \sup_{\delta > 0} \|\hat{m}_\delta\|_X,$$

where $X = K_1^{n(\frac{1}{p}-1),p}$ for $0 < p < 1$, $X = K(w)$ for $p = 1$, and a is a "unit p-atom" satisfying

$$
\begin{aligned}
&\operatorname{supp} a \subset \{|x| < 1\} , \\
(6.3) \qquad &|a(x)| \le 1 \\
&\int a(x) x^{\beta}\, dx = 0 , \qquad 0 \le |\beta| \le N .
\end{aligned}
$$

Take $\psi \in C_0^{\infty}$ with

$$\operatorname{supp} \psi \subset \{\tfrac{1}{2} \le |\xi| \le 2\} , \qquad 0 \le \psi \le 1 , \qquad \sum_{-\infty}^{\infty} \psi(\xi 2^{-j}) \equiv 1 .$$

For instance, we can let

$$\psi(\xi) = \frac{\bar{\psi}(\xi)}{\sum_{-\infty}^{\infty} \bar{\psi}(\xi 2^{-j})}$$

where $\bar{\psi} \in C_0^{\infty}$, $\operatorname{supp} \bar{\psi} \subset \{\frac{1}{2} \le |\xi| \le 2\}$, $0 \le \bar{\psi} \le 1$, $\bar{\psi} = 1$ on $\frac{5}{8} \le |\xi| \le \frac{15}{8}$.

Then $\psi\eta = \psi$ and

$$m(\xi)\, \hat{a}(\xi) = \sum_{-\infty}^{\infty} m(\xi)\, \eta(2^{-j}\xi)\, \hat{a}(\xi)\, \psi(2^{-j}\xi) = \sum_{-\infty}^{\infty} m_{2^j}(2^{-j}\xi)\, \hat{a}_{2^j}(2^{-j}\xi)$$

where $\hat{a}_{\delta}(\xi) = \hat{a}(\delta\xi)\, \psi(\xi)$. Write

$$f_j = (m_{2^j})^{\vee} , \qquad b_j = (\hat{a}_{2^j})^{\vee} , \qquad f = (m\hat{a})^{\vee} .$$

Then

$$(6.4) \qquad f(x) = \sum_{-\infty}^{\infty} 2^{nj} (f_j * b_j)(2^j x) .$$

We need some estimates for the b_j. Recall that a satisfies (6.3).

LEMMA 1. *Let* $r > 0$ *be given. Then*

(a) $$|b_j(x)| \leq C\, 2^{j(N+1)} (1+|x|)^{-r}, \quad \textit{if} \quad j \leq 0,$$

(b) $$|b_j(x)| \leq C\, 2^{-jn}, \quad \textit{all}\ x, \quad \textit{if} \quad j \geq 0,$$

$$|b_j(x)| \leq C\, |x|^{-r}, \quad |x| > 2^{j+1}, \quad \textit{if} \quad j \geq 0.$$

The constants C depend only on p, n, and r. This lemma, and the next one, will be proved at the end of the section.

To prove Theorem 3a we need another lemma.

LEMMA 2. *Suppose that* $0 < p < 1$, $j \geq 0$, $r > \frac{n}{p}$, $g \in K_1^{n(\frac{1}{p}-1),p}$, *and that* $Q \in L^1(R^n)$ *satisfies*

$$\int |Q|\, dx \leq 1, \qquad |Q(x)| \leq |x|^{-r} \quad \textit{if} \quad |x| > 2^{j+1}.$$

Then

$$\sum_{k=j+2}^{\infty} \Big(\int_{A_k} |g*Q|\, dx\Big)^p\, 2^{kn(1-p)} \leq C\, \|g\|^p_{K_1^{n(\frac{1}{p}-1),p}},$$

where C *depends only on* r, p, *and* n.

PROOF OF THEOREM 3a. Assume that $0 < p < 1$. We may assume that

(6.5) $$\sup_{\delta > 0} \|m_\delta\|_{K_1^{n(\frac{1}{p}-1),p}} = 1.$$

This implies that $\|\hat{m}_\delta\|_{L^1} \leq 1$ for every δ, and hence that $\|m\|_\infty \leq 1$ and also

$$(6.6) \qquad \|f_j\|_{L^1} \leq \|f_j\|_{K_1^{n(\frac{1}{p}-1),p}} \leq 1, \qquad -\infty < j < \infty .$$

We shall prove that when a satisfies (6.3)

$$(6.7) \qquad \|f\|_{K_1^{n(\frac{1}{p}-1),p}} \leq C .$$

Recall that $f = (m\hat{a})^{\vee}$. Since $K_1^{n(\frac{1}{p}-1),p}$ is continuously included in L^p, this will give (6.2) with L^p in place of H^p. The characterization of H^p in terms of iterated Riesz transforms will then enable us to finish the proof.

To prove (6.7), we first observe that

$$(6.8) \qquad \int_{|x|<4} |f|\, dx \leq C \left(\int |f|^2 dx\right)^{1/2} \leq C \left(\int |\hat{a}|^2 d\xi\right)^{1/2} \leq C$$

where we have used $\|m\|_\infty \leq 1$.

Next, assume $k \geq 2$. From (6.4) we have

$$(6.9) \qquad \int_{A_k} |f|\, dx \leq \sum_{j=-\infty}^{\infty} \int_{A_{k+j}} |f_j * b_j|\, dx = \sum_{j=-\infty}^{-k+1} + \sum_{j=-k+2}^{0} + \sum_{j=1}^{\infty}$$

$$\equiv \alpha_k + \beta_k + \gamma_k .$$

By (6.6) and Lemma 1 we have $\|f_j\|_{L^1} \leq 1$ and $\|b_j\|_{L^1} \leq C\, 2^{j(N+1)}$ for $j \leq 0$. Hence

$$\alpha_k \le \sum_{-\infty}^{-k+1} \|f_j * b_j\|_{L^1} \le C \sum_{-\infty}^{-k+1} 2^{j(N+1)} \le C\, 2^{-k(N+1)} ,$$

$$(6.10) \qquad \sum_{k=2}^{\infty} \alpha_k^p\, 2^{kn(1-p)} \le C \sum_{k=2}^{\infty} 2^{-kp(N+1) + kn(1-p)} \le C ,$$

since $N+1 > n(\frac{1}{p} - 1)$.

Next

$$\sum_{k=2}^{\infty} \beta_k^p\, 2^{kn(1-p)} \le \sum_{k=2}^{\infty} 2^{kn(1-p)} \sum_{j=-k+2}^{0} \Big(\int_{A_{k+j}} |f_j * b_j|\, dx\Big)^p$$

$$= \sum_{j=-\infty}^{0} \sum_{k=-j+2}^{\infty} 2^{kn(1-p)} \Big(\int_{A_{k+j}} |f_j * b_j|\, dx\Big)^p$$

$$= \sum_{j=-\infty}^{0} \sum_{\ell=2}^{\infty} 2^{n(\ell-j)(1-p)} \Big(\int_{A_\ell} |f_j * b_j|\Big)^p .$$

By lemma 1, the function $c\, 2^{-j(N+1)}\, b_j(x)$ satisfies the hypothesis of lemma 2 (for $j=0$) when $j \le 0$ and c is small enough. So, lemma 2 and (6.6) give

$$(6.11) \qquad \sum_{k=2}^{\infty} \beta_k^p\, 2^{kn(1-p)} \le C \sum_{j=-\infty}^{0} 2^{-nj(1-p)}\, 2^{jp(N+1)} \le C .$$

Similarly

$$\sum_{k=2}^{\infty} \gamma_k^p\, 2^{kn(1-p)} \le \sum_{j=1}^{\infty} \sum_{\ell=j+2}^{\infty} 2^{n(\ell-j)(1-p)} \Big(\int_{A_\ell} |f_j * b_j|\Big)^p .$$

By lemma 1, $c\, b_j(x)$ satisfies the hypothesis of Lemma 2 when c is small and $j \ge 0$. Lemma 2 and (6.6) give

$$(6.12)\qquad \sum_{k=2}^{\infty} \gamma_k^p \, 2^{kn(1-p)} \le \sum_{j=1}^{\infty} 2^{-j(1-p)n} \le C .$$

From (6.9) - (6.12) we see that

$$\sum_{k=2}^{\infty} \left(\int_{A_k} |f| \, dx\right)^p 2^{kn(1-p)} \le C$$

which, with (6.8), shows that (6.7) holds.

In particular, we have proved that when (6.5) holds

$$(6.13)\qquad \|(m\hat{a})^{\vee}\|_{L^p} \le C ,$$

for every p-atom a .

We want this inequality to hold with H^p norm instead of L^p norm. Fix a multi-index β and a number $\epsilon > 0$. Let R be the function with Fourier transform

$$\hat{R}(\xi) = \frac{\xi^{\beta}}{|\xi|^{|\beta|}} e^{-\epsilon|\xi|} .$$

Then $R*g$, $g \in L^2$, is a constant multiple of the Poisson integral of an iterated Riesz transform of g . Write $f = (m\hat{a})^{\vee}$ again .

We have

$$R*f = ((\hat{R}m)\hat{a})^{\vee} .$$

Write $\overline{m}(\xi) = \hat{R}(\xi)\, m(\xi)$. We claim that

$$(6.14)\qquad \sup_{\delta>0} \|(\overline{m}_{\delta})^{\wedge}\|_{K_1^{n(\frac{1}{p}-1),p}} \le C(\beta) \sup_{\delta>0} \|\hat{m}_{\delta}\|_{K_1^{n(\frac{1}{p}-1),p}} .$$

Then, using (6.5) and (6.13) applied to $\bar{m}$, it will follow that $\|R*f\|_{L^p} \le C(\beta)$. From the characterization of H^p in terms of iterated Riesz transforms [FS, §§8,11], it follows from this that $\|f\|_{H^p} \le C$ as required.

To prove (6.14), note that

$$\bar{m}_\delta(\xi) = \bar{m}(\delta\xi)\,\eta(\xi) = \hat{R}(\delta\xi)\,\varphi(\xi)\,m(\delta\xi)\,\eta(\xi)$$

where $\varphi \in C_0^\infty$ satisfies $\varphi = 1$ on $\frac{1}{4} \le |\xi| \le 4$, $\varphi = 0$ off $\frac{1}{8} \le |\xi| \le 8$. Thus

$$(\bar{m}_\delta)^\wedge = Q*\check{m}_\delta$$

where $\hat{Q}(\xi) = \varphi(\xi)\,\hat{R}(\delta\xi)$. For multi-indices γ we have $\|D^\gamma \hat{Q}\|_\infty \le C(\gamma, \beta)$, independently of ϵ and δ. So, for fixed $r > \frac{n}{p}$, there exists c such that cQ satisfies the hypothesis of Lemma 2 (with $j = 0$), for every δ and ϵ. (6.14) now follows from Lemma 2.

PROOF OF THEOREM 3B. We continue with the notation of the preceding proof. Now $p = 1$ and we shall assume $1 \le w(k)\uparrow$, $\sum_1^\infty w(k)^{-2} < \infty$, and normalize by

$$\sup_{\delta > 0} \|\hat{m}_\delta\|_{K(w)} = 1 . \tag{6.15}$$

Then

$$\|f_j\|_{L^1} \le \|f_j\|_{K(w)} \le 1 , \qquad -\infty < j < \infty . \tag{6.16}$$

The 1-atom a is assumed to satisfy (6.3). The estimate (6.8) follows as before, so we only need to show that

$$\int_{|x|>4} |f|\, dx \le C . \tag{6.17}$$

By (6.4) we have

$$f(x) = \sum_{j=-\infty}^{\infty} 2^{nj} (f_j * b_j)(2^j x) = \sum_{j=-\infty}^{0} + \sum_{j=1}^{\infty} = A(x) + B(x) .$$

Now

$$\int_{R^n} |A(x)|\, dx \le \sum_{j=-\infty}^{0} \|f_j * b_j\|_{L^1} \le C \sum_{j=-\infty}^{0} 2^{j(N+1)} \le C , \tag{6.18}$$

by (6.16) and Lemma 1.

Next,

$$\int_{|x|>4} |B(x)|\, dx \le \sum_{j=1}^{\infty} \int_{|x|>2^{j+2}} |f_j * b_j|\, dx \tag{6.19}$$

$$= \sum_{j=1}^{\infty} \sum_{k=j+2}^{\infty} \int_{A_k} |f_j * b_j| .$$

Fix j and k with $k \ge j+2$, $j \ge 1$, and suppose that $x \in A_k$. Write

$$f_j * b_j(x) = \int_{|y|<2^{k-1}} f_j(y)\, b_j(x-y)\, dy + (f_j \chi_{\bar{A}_k}) * b_j(x)$$

$$+ \int_{|y|>2^{k+2}} f_j(y)\, b_j(x-y)\, dy ,$$

where $\bar{A}_k = A_{k-1} \cup A_k \cup A_{k+1}$. Take $r = n+1$ in Lemma 1. Then

$$|b_j(x-y)| \le C|x|^{-(n+1)} \le C\,2^{-k(n+1)}\,, \qquad |y| < 2^{k-1}\,,$$

$$|b_j(x-y)| \le C|y|^{-(n+1)} \le C\,2^{-k(n+1)}\,, \qquad |y| > 2^{k+2}\,.$$

Hence, using (6.16),

$$|(f_j * b_j)(x)| \le C\,2^{-k(n+1)} + |f_j \chi_{\bar{A}_k}) * b_j(x)|\,,$$

and so

$$\int_{A_k} |f_j * b_j|\, dx \le C\,2^{-k} + \|f_j \chi_{\bar{A}_k}\|_{L^1} \|b_j\|_{L^1}$$

$$= C\,2^{-k} + C \sum_{\ell = k-1}^{k+1} J(\ell, j)\|b_j\|_{L^1}$$

where

$$J(k,j) = \int_{A_k} |f_j|\,.$$

Thus, by (6.19),

$$\int_{|x|>4} |B(x)|\, dx \le C + C \sum_{j=1}^{\infty} \sum_{k=j+1}^{\infty} J(k,j)\|b_j\|_{L^1}\,.$$

Since $w(k)\uparrow$, it follows from (6.16) that

$$\sum_{k=j+1}^{\infty} J(k,j) \le \frac{1}{w(j)} \sum_{k=j+1}^{\infty} w(k)\, J(k,j) \le \frac{1}{w(j)}\,, \qquad j \ge 0\,.$$

Hence

$$\int_{|x|>4} |B(x)|\, dx \le C + C \sum_{j=1}^{\infty} \frac{1}{w(j)} \|b_j\|_{L^1}$$

$$\le C + C \left(\sum_{j=1}^{\infty} \frac{1}{w(j)^2} \right)^{1/2} \left(\sum_{j=1}^{\infty} \|b_j\|_{L^1}^2 \right)^{1/2} .$$

We claim that

$$\sum_{j=1}^{\infty} \|b_j\|_{L^1}^2 \le C . \tag{6.20}$$

With the previous inequality and (6.18), (6.17), this will give for all atoms a and all m satisfying (6.15)

$$\|f\|_{L^1} \le C . \tag{6.21}$$

The proof of (6.20) is postponed temporarily. We will show now how to go from (6.21) to the H^1 inequality we need.

Let $\hat{R}(\xi) = \frac{\xi_j}{|\xi|} e^{-\epsilon|\xi|}$ where $\epsilon > 0$ and $j \in \{1,\dots,n\}$ is fixed. Then $\|R*a\|_{H^1} \le C$. By the atomic decomposition, there exist 1-atoms a_j and constants λ_j with $\Sigma|\lambda_j| \le C$ and

$$R*a = \sum_1^{\infty} \lambda_j a_j ,$$

with convergence in S'. Formally, we have

$$R*f = (\hat{R}(m\hat{a}))^{\vee} = (m(\hat{R}\hat{a}))^{\vee} = \Sigma \lambda_j (m\hat{a}_j)^{\vee} .$$

Since $m \in L^{\infty}$ and Fourier transformation is an isomorphism on S', the series on the right converges to $R*f$ in S'. But $\|(m\hat{a}_j)^{\vee}\|_{L^1} \leq C$, by (6.21), so in fact the series converges in L^1. Moreover

$$\|R*f\|_{L^1} \leq C \sum_1^{\infty} |\lambda_j| \leq C .$$

Hence $\|f\|_{H^1} \leq C$, by the Riesz transform characterization of H^1.

This completes the proof of Theorems 3a and 3b, modulo Lemmas 1 and 2 and (6.20).

PROOF OF (6.20). Recall that supp $a \subset (|x| < 1)$ and $|a(x)| \leq 1$. We have

$$(\int_{|x|<2^j} |b_j| \, dx)^2 \leq C\, 2^{nj} \int_{|x|<2^j} |b_j|^2 \leq C\, 2^{nj} \int |\hat{a}(2^j \xi)|^2 \, |\psi(\xi)|^2 \, d\xi$$

$$\leq C \int_{\bar{A}_j} |\hat{a}(\xi)|^2 \, d\xi$$

where $\bar{A}_j = A_{j-1} \cup A_j \cup A_{j+1}$. Hence

$$(6.22) \qquad \sum_{j=1}^{\infty} (\int_{|x|<2^j} |b_j| \, dx)^2 \leq C \int |\hat{a}(\xi)|^2 \, d\xi = C \int |a|^2 \, d\xi \leq C .$$

Also

$$(6.23) \qquad (\int_{|x|>2^j} |b_j| \, dx)^2 \leq C \, (\int |b_j|^2 \, |x|^{2n} \, dx)(\int_{|x|>2^j} |x|^{-2n} \, dx)$$

$$\leq C\, 2^{-jn} \sum_{i=1}^{n} \int |D_i^n \, \hat{b}_j|^2 \, d\xi ,$$

where $D_i^n = \dfrac{\partial^n}{\partial \xi_i^n}$.

Now $\hat{b}_j(\xi) = \hat{a}(2^j \xi)\, \psi(\xi)$, and since $j \geq 0$,

$$|(D_i^n \hat{b}_j)(\xi)| \leq C\, 2^{nj}\, \chi_{\frac{1}{2} \leq |\xi| \leq 2}(\xi) \sum_{\ell=0}^{n} |(D_i^{\ell} \hat{a})(2^j \xi)| \, .$$

Hence

$$\int |(D_i^n \hat{b}_j)|^2 \, d\xi \leq C\, 2^{nj} \sum_{\ell=0}^{n} \int_{\bar{A}_j} |D_i^{\ell} \hat{a}|^2 \, d\xi \, .$$

Combined with (6.23), this gives

$$\left(\int_{|x| > 2^j} |b_j| \, dx\right)^2 \leq C \sum_{i=1}^{n} \sum_{\ell=0}^{n} \int_{\bar{A}_j} |D_i^{\ell} \hat{a}|^2 \, d\xi \, ,$$

so that

$$\text{(6.24)} \qquad \sum_{j=1}^{\infty} \left(\int_{|x| > 2^j} |b_j| \, dx\right)^2 \leq C \sum_{i=1}^{n} \sum_{\ell=0}^{n} \int |D_i^{\ell} \hat{a}|^2 \, d\xi$$

$$= C \sum_{i=1}^{n} \sum_{\ell=0}^{n} \int |x_i|^{2\ell} \, |a(x)|^2 \, dx \leq C \, ,$$

which, with (6.22), gives (6.20).

PROOF OF LEMMA 1. We have

$$\text{(6.25)} \qquad b_j(x) = \int_{|y| < 1} a(y)\, \hat{\psi}(x - 2^j y)\, dy \, .$$

(a) Suppose that $j \leq 0$. Let $r > 0$ be given. For fixed x Taylor's formula gives

$$\hat{\psi}(x - 2^j y) = \sum_{|\beta| \leq N} C_\beta y^\beta + R(y) .$$

When $|y| < 1$, the facts that $\hat{\psi}$ has rapid decrease and $j \leq 0$ lead to the estimate

$$|R(y)| \leq C\, 2^{j(N+1)} |y|^{N+1} (1+|x|)^{-r} .$$

From (6.25) it follows that

$$|b_j(x)| = |\int_{|y|<1} a(y)\, R(y)\, dy| \leq C\, 2^{j(N+1)} (1+|x|)^{-r} ,$$

as required.

(b) Suppose that $j \geq 0$. Rewrite (6.25) as

$$b_j(x) = 2^{-nj} \int_{|y|<2^j} a(2^{-j} y)\, \hat{\psi}(x-y)\, dy .$$

Since $|a(x)| \leq 1$ and $\|\hat{\psi}\|_{L^1} \leq 1$ we have

$$|b_j(x)| \leq C\, 2^{-jn} , \qquad x \in R^n .$$

Suppose that $|x| > 2^{j+1}$. Then

$$|b_j(x)| \leq 2^{-jn} \int_{|y-x|<2^j} |a(2^{-j}(x-y))|\, |\hat{\psi}(y)|\, dy$$

$$\leq C\, 2^{-jn} \int_{|y| > \frac{1}{2}|x|} |\hat{\psi}(y)|\, dy$$

$$\leq C\, 2^{-jn} |x|^{-r} ,$$

since $\hat{\psi}$ has rapid decrease.

PROOF OF LEMMA 2. Fix $j \geq 0$. We may assume

$$\|g\|_{K_1^{n(\frac{1}{p}-1),p}} = 1 . \tag{6.26}$$

Suppose that $k \geq j+2$ and $x \in A_k$. Write

$$g*Q(x) = \sum_{\ell=-\infty}^{k-2} \int_{A_\ell} Q(x-y)\, g(y)\, dy + (g\, \chi_{\bar{A}_\ell}) * Q(x)$$

$$+ \sum_{\ell=k+2}^{\infty} \int_{A_\ell} Q(x-y)\, g(y)\, dy$$

where $\bar{A}_k = A_{k-1} \cup A_k \cup A_{k+1}$. If $x \in A_k$, $y \in A_\ell$, $\ell \leq k-2$, then $|Q(x-y)| \leq 2^{-r(k-1)}$, while if $\ell \geq k+2$ then $|Q(x-y)| \leq 2^{-\ell(k-1)}$.

Write $J(\ell) = \int_{A_\ell} |g|\, dx$. Then

$$|g*Q(x)| \leq 2^{-r(k-1)} \sum_{\ell=-\infty}^{k-2} J(\ell) + |(g\, \chi_{\bar{A}_k}) * Q(x)| + \sum_{\ell=k+2}^{\infty} 2^{-r(\ell-1)} J(\ell) ,$$

$$\int_{A_k} |g*Q|\, dx \leq C\, 2^{nk-rk} + \sum_{\ell=k-1}^{k+1} J(\ell) + C \sum_{\ell=k+2}^{\infty} 2^{nk-r\ell} J(\ell)$$

$$\leq C\, 2^{nk-rk} + C \sum_{\ell=k-1}^{\infty} J(\ell)\, 2^{nk-r\ell} ,$$

where we have used $\|g\|_{L^1} \leq \|g\|_{K_1^{n(\frac{1}{p}-1),p}} = 1$ and $r > \frac{n}{p} > n$. Hence, using (6.26),

$$\sum_{k=j+2}^{\infty} \left(\int_{A_k} |g*Q| \, dx\right)^p 2^{kn(1-p)}$$

$$\leq C \sum_{k=j+2}^{\infty} 2^{k(n-pr)} + C \sum_{\ell=j+1}^{\infty} \sum_{k=j+2}^{\ell+1} J(\ell)^p \, 2^{kn - r\ell p}$$

$$\leq C + C \sum_{\ell=j+1}^{\infty} J(\ell)^p \, 2^{\ell n - r\ell p}$$

$$\leq C + C \sum_{\ell=j+1}^{\infty} J(\ell)^p \, 2^{\ell n(1-p)} \leq C \, .$$

7. BEST POSSIBLE NATURE OF THEOREMS 3

We begin with two lemmas.

LEMMA 1. *Suppose that* $\alpha > 0$, $0 < q \leq \infty$, $r > n + \alpha$ *and that* $Q \in L^1(R^n)$ *satisfies*

(7.1) $$\|Q\|_{L^1} \leq 1 , \quad \text{and} \quad |Q(x)| \leq |x|^{-r} , \quad \text{for} \quad |x| > 2 .$$

Then, for $g \in K_2^{\alpha,q}$

$$\|g*Q\|_{K_2^{\alpha,q}} \leq C\,(\alpha,q,r,n)\,\|g\|_{K_2^{\alpha,q}} .$$

LEMMA 2. *Suppose that* $r > \frac{n}{p}$, $0 < p < 1$, Q *satisfies* (7.1) , *and* $\epsilon(k)$, $k \geq 1$, *satisfies* $\epsilon(k) \geq \epsilon(k+1) \geq 0$ *and*

(7.2) $$\epsilon(2k) \geq b\,\epsilon(k) , \quad \text{for some} \quad b > 0 .$$

Then

$$\|g*Q\|_{K(\epsilon,p)} \leq C(r,n,b,p)\,\|g\|_{K(\epsilon,p)} ,$$

where we define

$$\|g\|_{K(\epsilon,p)} = \|g\|_{L^1} + \sum_{k=0}^{\infty} \left(\int_{A_k} |g|\,dx\right)^p 2^{nk(1-p)}\,\epsilon(k) .$$

The proofs follow the lines of the proof of lemma 2 in §6. In the proof of this lemma 2 one also uses the inequalities

$$\varepsilon(k) \le C \left(\frac{\ell}{k}\right)^M \varepsilon(\ell) , \qquad 1 \le k \le \ell ,$$

for some $M > 0$, which follows from (7.2), and

$$\sum_{k=1}^{\ell} 2^{k\delta} \left(\frac{\ell}{k}\right)^M \le C(M,\delta)\, 2^{\delta\ell} , \qquad M > 0 , \quad \delta > 0 , \tag{7.3}$$

which follows from the inequalities $\frac{\ell}{k} \le \ell$ if $1 \le k \le \frac{1}{2}\ell$, $\frac{\ell}{k} \le 2$ if $\frac{1}{2}\ell \le k \le \ell$.

The following theorem illustrates the sharpness of Theorem 3a.

THEOREM 5a. *Suppose that* $0 < p < 1$ *and that* $\varepsilon(k)$, $k \ge 1$, *is a given sequence with* $0 < \varepsilon(k+1) \le \varepsilon(k)$. *There exist* m *and* $M \in L^\infty(R^n)$ *and* $h \in H^p$ *such that*

$$\sup_\delta \|\hat{m}_\delta\|_{K_2^{n(\frac{1}{p} - \frac{1}{2}),q}} < \infty , \qquad \textit{for every } q > p , \tag{7.4}$$

$$\sup_\delta \|\hat{M}_\delta\|_{K(\varepsilon,p)} < \infty , \tag{7.5}$$

but $(m\hat{h})^\vee \notin H^p$, *and* $(\hat{M}h)^\vee \notin H^p$.

By §2, an equivalent formulation of (7.4) is

$$\sup_\delta \|m_\delta\|_{B_2^{n(\frac{1}{p} - \frac{1}{2}),q}} < \infty ,$$

and by (1.6), the function m also satisfies

$$\sup_{\delta} \|\hat{m}_\delta\|_{K_1^{n(\frac{1}{p}-1),q}} < \infty , \qquad q > p .$$

PROOF OF THEOREM 5a. Define f_0 by

$$f_0(x) = 2^{-kn/p} k^{-1/p} , \qquad x \in A_k , \qquad k = 2,4,6,\ldots ,$$
$$= 0 , \qquad \text{otherwise} .$$

Then $f_0 \in \bigcap_{q>p} K_2^{n(\frac{1}{p}-\frac{1}{2}),q}$. Take $Q \in C^\infty$ satisfying $Q \geq 0$, $Q(0) > 0$, $\|Q\|_{L^1} = 1$, $\hat{Q} \in C^\infty$ and $\operatorname{supp} \hat{Q} \subset \{|x| \leq \frac{1}{2}\}$. For fixed $r > n(\frac{1}{p}+\frac{1}{2})$ Q satisfies $|Q(x)| \leq C\,|x|^{-r}$ for some C , and so cQ satisfies the hypothesis of Lemma 1 for some $c > 0$. Hence

$$f(x) \equiv e^{ix_1} (f_0 * Q)(x) \in \bigcap_{q>p} K_2^{n(\frac{1}{p}-\frac{1}{2}),q} .$$

Define $m = \hat{f}$. Then $\operatorname{supp} m \subset \{\frac{1}{2} \leq |\xi| \leq 2\}$ and $m_\delta(\xi) \equiv m(\delta\xi)\,\eta(\xi) = 0$ unless $\frac{1}{8} \leq \delta \leq 8$. The inverse Fourier transform of $m(\delta\xi)$ is $\delta^{-n} f(\delta^{-1}x)$. These functions form a bounded set in $K_2^{n(\frac{1}{p}-\frac{1}{2}),q}$ when $\frac{1}{8} \leq \delta \leq 8$. From Lemma 1 with $Q = c\check{\eta}$, it follows that m satisfies (7.4) .

Let $h = \check{\eta}$. Then

$$\int |h|^2 + |x|^{-2\alpha} |\hat{h}|^2 + |x|^{2\alpha} |h|^2)\, dx < \infty$$

for every $\alpha > 0$, so $h \in \bigcap_{p>0} H^p$, by Corollary 1 of §2.

We have

$$(m\hat{h})^{\vee} = \check{m} = f .$$

Since $f \in L^1$, to show $f \notin H^p$ it suffices to show $f \notin L^p$. Take $x \in A_k$, $k \geq 2$, k even. Then

$$(f_0 * Q)(x) = \int_{|y| < 2^{k-1}} Q(x-y)\, f_0(y)\, dy + \int_{A_k} Q(x-y)\, f_0(y)\, dy$$

$$+ \int_{|y| > 2^{k+2}} Q(x-y)\, f_0(y)\, dy .$$

Since $|Q(x)| \leq C\, |x|^{-r}$, the first and third integrals are $\leq C\, \|f_0\|_{L^1}\, 2^{-kr}$. Since $Q(0) > 0$ the middle integral is $\geq c\, 2^{-nk/p}\, k^{-1/p}$. Since $r > \frac{n}{p}$ it follows that for all sufficiently large k

$$(f_0 * Q)(x) \geq c\, 2^{-nk/p}\, k^{-1/p}, \qquad x \in A_k .$$

Hence $f_0 * Q \notin L^p$, and $f \notin L^p$, as required.

Next, we construct the function M. Take a subsequence $S \subset 2\,Z^+$ for which $\sum_{k \in S} \epsilon(k) < \infty$. Define

$$f_1(x) = 2^{-nk/p}, \qquad x \in A_k, \qquad k \in S,$$

$$= 0, \quad \text{otherwise}.$$

Let Q be as in the construction of m and define

$$g(x) = e^{ix_1}\, (f_1 * Q)(x), \qquad M = \hat{g} .$$

If $\epsilon(k)$ satisfies (7.2), then Lemma 2 and an analysis like the one above show that M satisfies (7.5) but that $(M\hat{h})^{\vee} \notin H^p$.

If $\epsilon(k)$ does not satisfy (7.2) we replace it with a sequence which does. We may suppose $\epsilon(1) = 1$. Define inductively integers ℓ_j, $j \geq 0$, by $\ell_0 = 0$, and ℓ_{j+1} is the smallest positive integer for which $\epsilon(2^{\ell_{j+1}}) \leq \frac{1}{2}\epsilon(2^{\ell_j})$. Define $\theta(k)$, $k \geq 1$, by

$$\theta(2^{\ell}) = 2^{-j}, \quad \text{if } \ell_j \leq \ell < \ell_{j+1}, \quad j \geq 0,$$

$$\theta(k) = \theta(2^{\ell}), \quad \text{if } 2^{\ell} \leq k < 2^{\ell+1}, \quad \ell \geq 0.$$

Then $\theta(k)\downarrow$ and $\theta(k)$ satisfies (7.2). Also, $\theta(k) \geq \epsilon(k)$. Construct M using $\theta(k)$. Then

$$\sup_{\delta>0} \|\hat{M}_\delta\|_{K(\epsilon,p)} \leq \sup_{\delta>0} \|\hat{M}_\delta\|_{K(\theta,p)} < \infty,$$

and the proof of Theorem 5a is complete.

Now we consider the case $p = 1$ and demonstrate the sharpness of Theorem 3b.

THEOREM 5b. *Suppose that* $w(k)$, $k \geq 1$, *satisfies*

$$(7.6) \qquad 1 \leq w(k)\uparrow \text{ as } k\uparrow, \qquad w(2k) \leq C\, w(k),$$

and $\sum_1^\infty w(k)^{-2} = \infty$. *Then there exist* $m \in L^\infty$ *and* $h \in H^1$ *such that*

$$\sup_{\delta>0} \|\hat{m}_\delta\|_{K(w)} < \infty$$

but $(m\hat{h})^\vee \notin H^1$.

Moreover,

$$\sup_{\delta>0} \|\hat{m}_\delta\|_{K_2^{\frac{n}{2},q}} < \infty, \qquad \text{for every } q > 0.$$

As we pointed out in §3, there exists a nondecreasing sequence $w(k)$ with $\Sigma w(k)^{-2} = \infty$ such that the only m satisfying $\sup_{\delta>0} \|\hat{m}_\delta\|_{K(w)} < \infty$ is $m \equiv 0$. Thus, some supplementary hypothesis such as $w(2k) \leq C w(k)$ is needed for the theorem to hold.

PROOF. Choose a set $S = \{k_1, k_2, k_3, \dots\} \subset Z^+$ with $k_i < k_{i+1}$, $\lim_{i\to\infty} (k_{i+1} - k_i) = \infty$, and

$$\sum_{k\in S} w(k)^{-2} = \infty .$$

We also assume that $k_1 \geq 10$ and $k_{i+1} - k_i \geq 10$ for $i \geq 1$. Take $\psi \in C_0^\infty$, $\psi \neq 0$, with $\operatorname{supp} \psi \subset \{|\xi| < 1\}$. Define

$$m(\xi) = \sum_{k\in S} \frac{1}{w(k)} \psi(2^k(\xi - 2^k e_1)) .$$

Then $m_\delta(\xi) = 0$ unless $2^{k-3} \leq \delta \leq 2^{k+3}$ for some $k \in S$. For fixed $k \in S$ and $2^{k-3} \leq \delta \leq 2^{k+3}$ we have

$$m_\delta(\xi) = \psi(2^k(\xi\delta - 2^k e_1)) \frac{1}{w(k)} \eta(\xi) .$$

Define $f_0 = \check{\psi}$, and fix $r > n$. Then f_0 satisfies

$$\|f_0\|_{L^1} \leq C , \qquad |f_0(x)| \leq C (1+|x|)^{-r} .$$

Write $\delta = 2^k \epsilon$, where $\frac{1}{8} \leq \epsilon \leq 8$. The inverse Fourier transform of $m(2^k(\xi\delta - 2^k e_1))$ is

$$F(x) = 2^{-2kn} \epsilon^{-n} f_0 (2^{-2k} \epsilon^{-1} x) e^{ix_1 \epsilon^{-1}} ,$$

which satisfies

$$(7.7) \qquad |F(x)| \le C\, 2^{-2kn}, \quad \text{all } x$$

$$|F(x)| \le C\, 2^{2k(r-n)}\, |x|^{-r}, \quad |x| > 2^{2k},$$

for some C which depends on r but not ϵ.

An argument like the one used to prove Lemma 2 of §6 shows that the functions $F*\hat{\eta}$ also satisfy (7.7). Since $\check{m}_\delta = \frac{1}{w(k)}(F*\hat{\eta})$, it follows that

$$(7.8) \qquad |\hat{m}_\delta(x)| \le C\, \frac{1}{w(k)}\, 2^{-2kn}, \quad \text{all } x,$$

$$|\hat{m}_\delta(x)| \le C\, \frac{1}{w(k)}\, 2^{2k(r-n)}\, |x|^{-r}, \quad |x| > 2^{2k}.$$

Hence

$$(7.9) \qquad \sum_{\ell=1}^{\infty} \left(\int_{A_\ell} |\hat{m}_\delta|\, dx\right) w(\ell) \le C \sum_{\ell=1}^{2k} \frac{w(\ell)}{w(k)}\, 2^{-2kn}\, 2^{n\ell}$$

$$+ C \sum_{\ell=2k+1}^{\infty} \frac{w(\ell)}{w(k)}\, 2^{2k(r-n)}\, 2^{(n-r)\ell}.$$

The hypotheses (7.6) imply that $w(\ell) \le C\,w(k)$ for $1 \le \ell \le 2k$ and also

$$(7.10) \qquad w(\ell) \le C \left(\frac{\ell}{k}\right)^M w(k), \quad \ell > k,$$

for some $C > 0$. Thus, the right hand side of (7.9) is majorized by

$$C + C\, 2^{2k(r-n)} \sum_{\ell=2k}^{\infty} \left(\frac{\ell}{k}\right)^M 2^{(n-r)\ell}.$$

The last sum is easily shown to be $O(2^{(n-r)2k})$. Hence, the left hand side of (7.9) is $O(1)$, uniformly in δ, and we have shown that

$$\sup_{\delta>0} \|\hat{m}_\delta\|_{K(w)} < \infty .$$

Also, from (7.8) it follows that

$$\sup_{\delta>0} \|\hat{m}_\delta\|_{K_2^{\frac{n}{2},q}} < \infty , \qquad \text{for every } q>0 ,$$

since $w(k) \geq 1$.

It remains to construct the function h. Choose $k_1 < k_2 < \ldots$ in S such that

$$1 \leq \sum_{k\in S(j)} w(k)^{-2} < 2 , \qquad j = 1,2, \ldots$$

where $S(j) = \{k \in S : k_j \leq k < k_{j+1}\}$.

Define $\varphi, \lambda : S \to Z^+$ by

$$\varphi(k) = j^{-1} , \qquad \text{if } k \in S(j) ,$$

$$\lambda(k) = \frac{\varphi(k)}{w(k)} .$$

Then

$$\sum_{k\in S} \lambda(k)^2 < \infty , \qquad \sum_{k\in S} \frac{\lambda(k)}{w(k)} = \infty .$$

Take $g_0 \in C_0^\infty$ with $\operatorname{supp} g_0 \subset \{|\xi| < 1\}$, $g_0 = 1$ on $|\xi| < \frac{1}{2}$.

Define

$$g(\xi) = \sum_{k\in S} \lambda(k)\, g_0\,(\xi - 2^k e_1)\,, \qquad h = \check{g}\,.$$

Then

$$\int (|g|^2 + |x|^{-2\alpha}\,|g|^2 + |x|^{2\alpha}\,|\hat{g}|^2)\,dx < \infty$$

for every $\alpha > 0$, so $h \in H^1$, by Corollary 1 of §2.

We have

$$m(\xi)\,g(\xi) = \sum_{k\in S} \frac{\lambda(k)}{w(k)}\,\psi(2^k(\xi - 2^k e_1))\,,$$

$$f(x) \equiv (mg)^{\vee}(x) = \sum_{\ell\in S} \frac{\lambda(\ell)}{w(\ell)}\, f_0\,(x\, 2^{-\ell})\, 2^{-n\ell}\, e^{ix_1 2^{\ell}}\,,$$

where $f_0 = \hat{\psi}$. Fix $k \in S$. Write

$$f(x) = \sum_{\substack{\ell\in S\\ \ell<k}} + \sum_{\substack{\ell\in S\\ \ell=k}} + \sum_{\substack{\ell\in S\\ \ell>k}} = f_1(x) + f_2(x) + f_3(x)\,. \tag{7.11}$$

Notice that

$$\int_{A_k} |f_2|\,dx = \frac{\lambda(k)}{w(k)} \int_{A_0} |f_0|\,dx = c\,\frac{\lambda(k)}{w(k)}\,. \tag{7.12}$$

Here $c > 0$ since $\psi \in C_0^\infty$ and $\psi \neq 0$.

Next, since $|f_0(x)| \leq C$,

$$|f_3(x)| \leq C \sum_{\substack{\ell>k\\ \ell\in S}} \frac{\lambda(\ell)}{w(\ell)}\, 2^{-n\ell} \leq C\,\frac{\lambda(k)}{w(k)}\, 2^{-nk''}\,,$$

where k'' denotes the successor of k in S. Hence

(7.13)
$$\int_{A_k} |f_3|\, dx \leq C \frac{\lambda(k)}{w(k)} 2^{n(k-k'')} .$$

Finally, since $|f_0(x)| \leq C|x|^{-r}$ for $|x|>1$, if $x \in A_k$ then

(7.14)
$$|f_1(x)| \leq C \sum_{\substack{\ell \in S \\ \ell < k}} \frac{\lambda(\ell)}{w(\ell)} 2^{-r(k-\ell)} 2^{-n\ell} .$$

We claim that, for any $k \in S$ and $\delta > 0$,

(7.15)
$$\sum_{\substack{\ell \in S \\ \ell \leq k}} \frac{\lambda(\ell)}{w(\ell)} 2^{\delta\ell} \leq C(\delta) \frac{\lambda(k)}{w(k)} 2^{\delta k} .$$

Assume (7.15) for the moment. Return to (7.14) and let k' be the predecessor of k in S. Then, with $\delta = r - n$ and k' in place of k

$$|f_1(x)| \leq C \frac{\lambda(k')}{w(k')} 2^{-rk} 2^{(r-n)k'} , \qquad x \in A_k .$$

Hence

(7.16)
$$\int_{A_k} |f_1|\, dx \leq C \frac{\lambda(k')}{w(k')} 2^{(n-r)(k-k')} .$$

Using (7.10) and the fact that $k_{i+1} - k_i \to \infty$ for $k_i \in S$, it is easy to show that the right hand side of (7.16) is $o(\frac{\lambda(k)}{w(k)})$. From this and (7.13) it follows that

$$\int_{A_k} (|f_1| + |f_3|)\, dx = o(\frac{\lambda(k)}{w(k)}) , \quad k \to \infty .$$

So, from (7.11) and (7.12) it follows that

$$\int_{A_k} |f| \geq c \frac{\lambda(k)}{w(k)} , \qquad k \in S , \quad k \text{ large,}$$

for some $c > 0$. Since

$$\sum_{k \in S} \frac{\lambda(k)}{w(k)} = \infty ,$$

we see that $f \notin L^1$, and therefore

$$f = (m\hat{h})^{\vee} \notin H^1 .$$

PROOF OF (7.15). If $\varphi(k) = 1$ then $\lambda(\ell) = w(\ell)^{-1}$ for $\ell \leq k$ and (7.15) follows from (7.10) and (7.3). Assume $\varphi(k) \leq \frac{1}{2}$. Let k_0 be the largest element of S such that

$$\varphi(k_0) \geq 2\varphi(k) .$$

Since $\varphi(k)^{-1} - \varphi(k_0)^{-1} \leq (k - k_0)$, we have

$$\varphi(k)^{-1} \leq \varphi(k_0)^{-1} + (k - k_0) \leq \frac{1}{2}\varphi(k)^{-1} + (k - k_0)$$

so that for some $C = C(M, \delta)$

$$\varphi(k)^{-1} \leq 2\,(k - k_0) \leq C \left(\frac{k_0}{k}\right)^{2M} 2^{\delta(k - k_0)} . \tag{7.17}$$

Here M is the number in (7.10). From (7.17) follows

$$\frac{\varphi(k)}{w(k)^2} \geq C\, w(k)^{-2} \left(\frac{k}{k_0}\right)^{2M} 2^{\delta(k_0 - k)} . \tag{7.18}$$

Using $\varphi \leq 1$, (7.10) and (7.3), we obtain

$$\sum_{\substack{\ell \in S \\ \ell \leq k}} \frac{\lambda(\ell)}{w(\ell)} 2^{\delta\ell} = \sum_{\substack{\ell \in S \\ \ell \leq k}} \frac{\varphi(\ell)}{w(\ell)^2} 2^{\delta\ell} = \sum_{\ell \leq k_0} + \sum_{\ell = k_0+1}^{k}$$

$$\leq C\, w(k_0)^{-2}\, 2^{\delta k_0} + C\varphi(k)\, w(k)^{-2}\, 2^{\delta k}$$

$$\leq C\, w(k)^{-2} \left(\frac{k}{k_0}\right)^{2M} 2^{\delta k_0} + C\,\varphi(k)\, w(k)^{-2}\, 2^{\delta k} .$$

By (7.18), the first term in the last expression is $\leq \varphi(k)\, w(k)^{-2}\, 2^{\delta k} = \lambda(k)\, w(k)^{-1}\, 2^{\delta k}$, and the proof of (7.15) is complete.

8. LOWER MAJORANT THEOREM

We shall prove the following result.

THEOREM 6. *For each* $f \in H^p$ *there exists* $g \in H^p$ *such that*

$$|\hat{f}(\xi)| \le \hat{g}(\xi) \quad \text{and} \quad \|g\|_{H^p} \le C(p,n)\, \|f\|_{H^p} .$$

For $p=1$, $n=1$ this is a theorem of Hardy-Littlewood in the periodic case (See [Z, p. 287]). Coifman and Weiss [CW, p. 584] used the atomic decomposition to give a new proof for $H^1(R)$. Weiss [W, p. 199] raised the question of whether this "lower majorant property" would hold for $H^1(R^n)$.

Theorem 6, which states that $H^p(R^n)$ has the lower majorant property for every $p \in (0,1]$, was proved by the authors during the early stages of this project (See Abstracts of the AMS, 1(1980), p. 444). A similar proof was found independently by A.B. Alexandrov [A].

The analogous problem for L^p, $1 \le p < \infty$, has an interesting history. The interested reader is referred to [LS].

PROOF OF THEOREM 6. By the atomic decomposition (see §6) it suffices to consider the case when f is a unit p-atom, that is f has vanishing moments up to order N and satisfies

$$\text{(8.1)} \qquad \operatorname{supp} f \subset \{|x| \le 1\}, \qquad \|f\|_\infty \le 1 .$$

Let $Q = [0,1]^n$ denote the closed unit cube. For $\ell \in Z^n$ define

$$a(\ell) = \sup_{\xi \in Q} |\hat{f}(\xi + \ell)| .$$

The key property of $\hat{f}$ is the inequality

$$\sum_{\ell \in Z^n} |a(\ell)|^2 \leq C . \tag{8.2}$$

This inequality has been discovered independently by (at least) Sledd-Stegenga [SS], J. Stewart [St], Alexandrov [A] and the present authors.

We also need the fact that (8.1) implies

$$|\hat{f}(\xi)| \leq C |\xi|^{N+1} . \tag{8.3}$$

To see this, note that $\hat{f}$ is an entire function of n complex variables. Expand it in Taylor series around the origin

$$\hat{f}(\xi) = \Sigma\, c_\beta\, \xi^\beta .$$

The vanishing moments condition shows that $c_\beta = 0$ for $|\beta| \leq N$, while (8.1) implies that $|c_\beta| \leq \frac{C}{(\beta_1)!\ldots(\beta_n)!}$ and (8.3) easily follows.

Take $\varphi \in C_0^\infty$ with $\varphi = 1$ on Q, $\operatorname{supp} \varphi \subset \frac{3}{2} Q$ (the cube with center $(\frac{1}{2},\ldots,\frac{1}{2})$ and sidelength $\frac{3}{2}$). Define

$$G_1(\xi) = \Sigma\, a(\ell)\, \varphi(\xi + \ell)$$

where the sum is taken over all $\ell \in Z^n$ for which $Q + \ell$ does not contain the origin. Take $\psi \in C_0^\infty$ such that $\psi = 1$ on all the cubes $Q + \ell$ which do contain the origin. Let

$$G_2(\xi) = C\, |\xi|^{N+1}\, \psi(\xi) ,$$

where this C is the same one as in (8.3). Then

$$|\hat{f}(\xi)| \leq G_1(\xi) + G_2(\xi), \qquad \xi \in R^n.$$

From (8.2) it follows that

$$\int |G_1(\xi)|^2 (1+|\xi|^{-2\alpha})\, d\xi + \int |\hat{G}_1(x)|^2 |x|^{2\alpha}\, dx \leq C(\alpha) \tag{8.4}$$

for every $\alpha > 0$, and hence $\|\hat{G}_1\|_{H^p} \leq C$, by Corollary 1 of §2.

If N is odd then $G_2 \in C_0^\infty$ and (8.4) holds with G_2 in place of G_1 for $\alpha < N+1+\frac{1}{2}n$. Choosing $\alpha \in (n(\frac{1}{p}-\frac{1}{2}), N+1+\frac{1}{2}n)$, it follows as above that $\|\hat{G}_2\|_{H^p}$. Thus, the conclusion of Theorem 6 holds, with $g = \hat{G}_1 + \hat{G}_2$.

In case N is even, one can use standard methods of Fourier analysis, like those in [S, chapter 3], to prove that

$$\hat{G}_2(x) = O(|x|^{-(n+N+1)}), \quad x \to \infty, \tag{8.5}$$

so that (8.4) again holds for $\alpha < N+1+\frac{1}{2}n$ and the proof of Theorem 6 follows as in the case N odd.

Here is another proof for the case N even, which avoids (8.5). Let A be a constant such that

$$|\xi|^{N+2} \leq A \sum_{i=1}^{n} \xi_i^{N+2}.$$

Define, with C the constant of (8.3)

$$G_3(\xi) = AC \sum_{i=1}^{n} \frac{\xi_i}{|\xi|} \xi_i^{N+1} \psi(\xi).$$

Then $|\hat{f}(\xi)| \leq G_1(\xi) + G_3(\xi)$. Since $\xi_i^{N+1} \psi \in C_0^\infty$, it follows as above that

$$\|(\xi_i^{N+1} \psi)^\wedge\|_{H^p} \leq C$$

for $i = 1, \ldots, n$. Since Riesz transforms are bounded on H^p it follows that $\|\hat{G}_3\|_{H^p} \leq C$, so that the conclusion of Theorem 6 holds with $g = \check{G}_1 + \check{G}_3$.

9. ON A THEOREM OF PIGNO AND SMITH

Suppose $q > 1$, and let $\{b_j\}_1^\infty$ be a sequence of positive numbers satisfying $b_1 = 1$ and

$$(9.1) \qquad \frac{b_{j+1}}{b_j} \geq q , \qquad j \in Z^+ .$$

Pigno and Smith [PS 1], see also [PS 2], assumed that $q \geq 2$ and used the method of Cohen-Davenport to prove the following theorem about $H^1(\mathbb{T})$: Given an analytic function $f \in H^1(\mathbb{T})$ there are measures $\mu_j \in M(\mathbb{T})$ satisfying

$$\hat{\mu}_j(\ell) = \hat{f}(\ell) , \qquad b_j \leq \ell \leq b_{j+1} , \qquad j = 1, 2, \ldots , \qquad \ell \in Z ,$$

and

$$\sum_{j=1}^{\infty} \|\mu_j\|_{M(\mathbb{T})}^2 \leq C\|f\|_{H^1(\mathbb{T})}^2 .$$

We are going to use the atomic decomposition to prove an analogous result for $H^p(\mathbb{R}^n)$, $0 < p \leq 1$. Suppose that $q > 1$ and that $\{b_j\}_{-\infty}^\infty$ is a strictly increasing sequence of positive numbers satisfying (9.1) for every $j \in Z$.

THEOREM 7. *For each* $f \in H^p(\mathbb{R}^n)$, $0 < p \leq 1$, $n \geq 1$, *there exists a sequence* $\{f_j\}_{-\infty}^\infty$ *in* $H^p(\mathbb{R}^n)$ *such that*

$$\hat{f}_j = \hat{f} \text{ on } b_j \leq |\xi| \leq b_{j+1} , \qquad j \in Z ,$$

and

$$\sum_{j=-\infty}^{\infty} \|f_j\|_{H^p}^2 \leq C\|f\|_{H^p}^2 ,$$

where $C = C(p,q,n)$ *depends on* q *but not on* $\{b_j\}$.

Examples of the form $\sum_{k=1}^{\infty} \lambda_k \varphi(\xi - 2^k e_1)$, where $\varphi \in C_0^\infty$, $\operatorname{supp} \varphi \subset \{|\xi| < 1\}$, $\Sigma \lambda_k^2 < \infty$, which belong to $\bigcap_{0<p\leq 1} H^p$ by Corollary 1 of §2 , suggest that $\|f_j\|_{H^p}^2$ cannot be replaced by $\|f_j\|_{H^p}^r$ for any $r < 2$.

PROOF. An easy argument shows it suffices to prove the theorem when f is a unit p-atom, that is, one satisfying (8.1) .

Let j_0 be the integer such that

$$b_{j_0} \leq 1 < b_{j_0+1} ,$$

and write

$$B_j = \{b_j - \tfrac{1}{2} \leq |\xi| \leq b_{j+1} + \tfrac{1}{2}\} .$$

For $j > j_0$ choose $\varphi_j \in C_0^\infty$ satisfying $\varphi_j = 1$ on $b_j \leq |\xi| \leq b_{j+1}$, $\operatorname{supp} \varphi_j \subset B_j$, and

$$\|D^\beta \varphi_j\|_\infty \leq C(\beta, n) , \tag{9.2}$$

where $C(\beta,n)$ does not depend on the b_j .

Define $F_j = \varphi_j \hat{f}$ and let $f_j = \check{F}_j$. Take $s \in Z^+$ with $s > n(\frac{1}{p} - \frac{1}{2})$. From (9.2) and Corollary 1 of §2 it follows that

$$\|f_j\|^2_{H^p} \leq C\int|F_j|^2 + C \sum_{|\beta|=s} \int|D^\beta F_j|^2 \, d\xi$$

$$\leq C\int_{B_j} |\hat{f}|^2 + C \sum_{|\beta|\leq s} \int_{B_j} |D^\beta \hat{f}|^2 \, d\xi .$$

Since $b_{j+1} \geq q\, b_j$, each ξ with $|\xi| > \frac{1}{2}$ can belong to at most $C(q)$ of the B_j. Hence

$$\sum_{j=j_0+1}^{\infty} \|f_j\|^2_{H^p} \leq C\int|\hat{f}|^2 + C \sum_{|\beta|\leq s} \int|D^\beta \hat{f}|^2 \tag{9.3}$$

$$\leq C\int|f|^2 + C \sum_{|\beta|\leq s} \int|x^\beta f(x)|^2 \, dx \leq C .$$

Next we consider the case $j \leq j_0 - 1$. Choose $\psi \in C_0^\infty$ with $\psi = 1$ on $|\xi| < 1$, $\operatorname{supp} \psi \subset \{|\xi| < 2\}$. Define

$$F_j(\xi) = \hat{f}(\xi)\, \psi(b_{j+1}^{-1} \xi) , \qquad j \leq j_0 - 1 ,$$

and let $f_j = \check{F}_j$. The proof of Lemma 1a in §6 uses only that $\psi \in S$ and shows that if $\delta \leq 1$ the Fourier transform h of $\hat{f}(\delta\xi)\, \psi(\xi)$ satisfies

$$|h(x)| \leq C(r)(1+|x|)^{-r} \delta^{N+1} ,$$

for any $r > 0$

Also, since by (8.3) $|\hat{f}(\xi)| \leq C|\xi|^{N+1}$, we have

$$|\hat{f}(\delta\xi)\, \psi(\xi)| \leq C\delta^{N+1} |\xi|^{N+1}$$

Choosing $\alpha \in (n(\frac{1}{p} - \frac{1}{2}) , N+1+\frac{n}{2})$ and $r > \alpha + n$, Corollary 1 of §2 shows that $\|h\|_{H^p} \leq C\delta^{N+1}$.

Letting $\delta = b_{j+1}$, we have

$$f_j(x) = b_{j+1}^n \, h(b_{j+1} x) \, , \qquad \|f_j\|_{H^p} = b_{j+1}^{n(1-\frac{1}{p})} \|h\|_{H^p} \leq C \, b_{j+1}^{\varepsilon} \, ,$$

where $\varepsilon = N+1-n(\frac{1}{p}-1) > 0$. Hence

$$(9.4) \qquad \sum_{j=-\infty}^{j_0-1} \|f_j\|_{H^p}^2 \leq C \sum_{j=-\infty}^{j_0} b_j^{2\varepsilon} \leq C \sum_{j=-\infty}^{j_0} (b_{j_0} q^{j-j_0})^{2\varepsilon}$$

$$\leq C \, b_{j_0}^{2\varepsilon} \leq C \, .$$

Theorem 7 follows from (9.3) , (9.4) , and the choice $f_{j_0} = f$.

10. EXTENSION OF A THEOREM OF OBERLIN

A theorem of D.M. Oberlin [O] asserts that for $f \in H^1(\mathbb{R}^n)$,

$$\sum_{k=-\infty}^{\infty} \sup_{2^k \le r \le 2^{k+1}} \int_{|\xi|=1} |\hat{f}(r\xi)| \, d\sigma(\xi) \le C\|f\|_{H^1}, \qquad n \ge 2,$$

where $d\sigma$ is surface measure on $|x| = 1$.

We will extend this result in two ways - by replacing the $n-1$ dimensional spheres with circles and the L^1 norms by L^q norms, $1 \le q < \infty$.

THEOREM 8. _Suppose that_ $1 \le q < \infty$ _and_ $n \ge 2$. _There exist constants_ $C(n,q)$ _with the following property: Let_ S_k, $k \in \mathbb{Z}$, _be any sequence of circles in_ $\mathbb{R}^n$ _with center at the origin and radius between_ 2^k _and_ 2^{k+1}. _Then_

$$\sum_{k=-\infty}^{\infty} \left(\int_{S_k} |\hat{f}(\xi)|^q \frac{|d\xi|}{|\xi|}\right)^{1/q} \le C(q,n)\|f\|_{H^1}. \tag{10.1}$$

As a corollary, we obtain the inequality

$$\sum_{k=-\infty}^{\infty} \sup_{2^k \le r < 2^{k+1}} \left(\int_{|\xi|=1} |\hat{f}(r\xi)|^q \, d\sigma(\xi)\right)^{1/q} \le C(q,n)\|f\|_{H^1}.$$

The theorem and corollary are false for $q = \infty$. A counterexample is furnished by

$$\hat{f}(\xi) = \sum_{k=1}^{\infty} \frac{1}{k} \varphi(\xi - 2^k e_1)$$

where $\varphi \in C_0^\infty$, $\operatorname{supp} \varphi \subset (|\xi| < 1)$, $\varphi(0) = 1$. Corollary 1 of §2 shows that $f \in H^1$.

For $p < 1$ a proof like the one of Theorem 8 leads to the inequality

$$\sum_{k=-\infty}^{\infty} 2^{-kn(1-p)} \left(\int_{S_k} |\hat{f}|^q \frac{|d\xi|}{|\xi|}\right)^{p/q} \le C(n,q,p) \|f\|_{H^p}^p .$$

But here the case $q = \infty$ holds too,

$$\sum_{k=-\infty}^{\infty} 2^{-kn(1-p)} \sup_{\xi \in A_k} |\hat{f}(\xi)|^p \le C(n,p) \|f\|_{H^p}^p . \tag{10.2}$$

The proof of (10.2) is left to the reader. It extends the inequality $|\hat{f}(\xi)| \le C|\xi|^{n(\frac{1}{p}-1)} \|f\|_{H^p}$ which we have used earlier.

PROOF OF THEOREM 8. We may assume f is a unit 1-atom, so that (8.1) is satisfied. By Hölder's inequality, we may also assume $q \ge 2$. By (8.3) we have $|\hat{f}(\xi)| \le C|\xi|$, so that

$$\sum_{k=-\infty}^{0} \left(\int_{S_k} |\hat{f}(\xi)|^q \frac{|d\xi|}{|\xi|}\right)^{1/q} \le C . \tag{10.3}$$

To analyze the case $k \ge 1$ we introduce the unit cube Q and the numbers

$$a(\ell) = \sup_{\xi \in Q} |\hat{f}(\xi + \ell)| , \qquad \ell \in Z^n ,$$

as in §8. Let $\mathcal{L}_k = \{\ell \in Z^n : Q + \ell \text{ meets } S_k\}$. Each intersection has arc length $\le C(n)$, so

$$\int_{S_k} |\hat{f}|^q \frac{|d\xi|}{|\xi|} \leq C2^{-k} \sum_{\ell \in \mathcal{L}_k} a(\ell)^q \ .$$

By Hölder's inequality, with $\frac{1}{q} + \frac{1}{q'} = 1$,

$$\sum_{k=1}^{\infty} \left(\int_{S_k} |\hat{f}|^q \frac{|d\xi|}{|\xi|}\right)^{1/q} \leq C \sum_{k=1}^{\infty} 2^{-k/q} \left(\sum_{\ell \in \mathcal{L}_k} a(\ell)^q\right)^{1/q}$$

$$\leq C\left(\sum_{k=1}^{\infty} 2^{-kq'/q}\right)^{1/q'} \left(\sum_{k=1}^{\infty} \sum_{\ell \in \mathcal{L}_k} a(\ell)^q\right)^{1/q} \ .$$

Now each ℓ belongs to at most two of the $\mathcal{L}_k$, and we are assuming $q \geq 2$. Hence

$$\sum_{k=1}^{\infty} \left(\int_{S_k} |\hat{f}|^q \frac{|d\xi|}{|\xi|}\right)^{1/q} \leq C\left(\sum_{\ell \in Z^n} a(\ell)^2\right)^{1/2} \leq C \ ,$$

where we have used (8.2) . This inequality, with (10.3), proves Theorem 8 .

REFERENCES

[A] A.B. Alexandrov, A majorization property for the several variable Hardy-Stein-Weiss classes (in Russian) Vestnik Leningrad. Univ. No. 13 (1982), 97-98.

[C] R.R. Coifman, A real variable characterization of H^p, Studia Math. 51 (1974), 269-274.

[CW] R.R. Coifman and G. Weiss, Extensions of Hardy spaces and their use in analysis, Bull. Amer. Math. Soc. 83 (1977), 569-645.

[CT] A.P. Calderón and A. Torchinsky, Parabolic maximal functions associated with a distribution, II, Advances in Math. 24 (1977), 101-171.

[F] T.M. Flett, Some elementary inequalities for integrals with applications to Fourier transforms, Proc. London Math. Soc. (3) 29 (1974), 538-556.

[FS] C. Fefferman and E.M. Stein, H^p spaces of several variables, Acta Math. 129 (1972), 137-193.

[H] C. Herz, Lipschitz spaces and Bernstein's theorem on absolutely convergent Fourier transforms, J. Math. Mech. 18 (1968), 283-324.

[Ja] S. Janson, Generalizations of Lipschitz spaces and an application to Hardy spaces and bounded mean oscillation, Duke Math J. 47 (1980), 959-982.

[Jo 1] R. Johnson, Temperatures, Riesz potentials, and the Lipschitz spaces of Herz, Proc. London Math. Soc. (3) 27 (1973) 290-316.

[Jo 2] R. Johnson, Multipliers of H^p spaces, Ark. Mat. 16 (1977), 235-249.

[L] R.H. Latter, A characterization of $H^p(R^n)$ in terms of atoms, Studia Math. 62 (1978), 93-101.

[LS] E.T.Y. Lee and G-I. Sunouchi, On the majorant properties in $L^p(G)$, Tôhoku Math. J. 31 (1979), 41-48.

[MS] S. Minakshisundaram and O. Szasz, On absolute convergence of Fourier series, Trans. Amer. Math. Soc. 61 (1947), 36-53.

[M] A. Miyachi, On some Fourier multipliers for $H^p(R^n)$; J. Fac. Sci. Tokyo 27 (1980), 157-179.

[O] D.M. Oberlin, A multiplier theorem for $H^1(R^n)$, Proc. Amer. Math. Soc. 73 (1979), 83-88.

[P] J. Peetre, New Thoughts on Besov Spaces, Duke University Mathematics Dept., Durham, 1976.

[PT] J. Peral and A. Torchinsky, Multipliers in $H^p(R^n)$, $0<p<\infty$, Ark. Mat. 17 (1978), 225-235.

[PS 1] L. Pigno and B. Smith, A Littlewood-Paley inequality for analytic measures, Ark. Mat. 20 (1982), 271-274.

[PS 2] L. Pigno and B. Smith, Quantitative behavior of the norms of an analytic measure, Proc. Amer. Math. Soc. 86 (1982), 581-585.

[Sj] P. Sjölin, An H^p inequality for strongly singular integrals, Math Z. 165 (1979), 231-238.

[SS] W. Sledd and D. Stegenga, An H^1 multiplier theorem, Ark. Mat. 19 (1981), 265-270.

[S] E.M. Stein, Singular Integrals and Differentiability Properties of Functions, Princeton Univ. Press, Princeton, 1970.

[Stw] J. Stewart, Fourier transforms of unbounded measures, Can. J. Math. 31 (1979), 1281-1292.

[Sz] O. Szasz, Fourier series and mean moduli of continuity, Trans. Amer. Math. Soc. 42 (1937), 366-395.

[T] M. Taibleson, On the theory of Lipschitz spaces of distributions on Euclidean n-space, I, J. Math. Mech. 13 (1964), 407-480; II, ibid 14 (1965), 821-840; III, ibid 15 (1966), 973-981.

[TW] M. Taibleson and G. Weiss, The molecular characterization of certain Hardy spaces, Astérisque 77 (1980), 67-151.

[W] G. Weiss, Some problems in the theory of Hardy spaces, Proc. Symp. Pure Math. 35, A.M.S., Providence, 1979.

[Wi] J.M. Wilson, A simple proof of the atomic decomposition for $H^p(R^n)$, $0<p\leq 1$, Studia Math. 74 (1982), 25-33.

[Z] A. Zygmund, Trigonometric Series, 2nd ed., Cambridge University Press, Cambridge, 1968.

Washington University
St. Louis, Missouri 63130

McMaster University
Hamilton, Ontario L8S 4K1

General instructions to authors for
PREPARING REPRODUCTION COPY FOR MEMOIRS

For more detailed instructions send for AMS booklet, "A Guide for Authors of Memoirs." Write to Editorial Offices, American Mathematical Society, P. O. Box 6248, Providence, R. I. 02940.

MEMOIRS are printed by photo-offset from camera copy fully prepared by the author. This means that, except for a reduction in size of 20 to 30%, the finished book will look exactly like the copy submitted. Thus the author will want to use a good quality typewriter with a new, medium-inked black ribbon, and submit clean copy on the appropriate model paper.

Model Paper, provided at no cost by the AMS, is paper marked with blue lines that confine the copy to the appropriate size. Author should specify, when ordering, whether typewriter to be used has PICA-size (10 characters to the inch) or ELITE-size type (12 characters to the inch).

Line Spacing – For best appearance, and economy, a typewriter equipped with a half-space ratchet – 12 notches to the inch – should be used. (This may be purchased and attached at small cost.) Three notches make the desired spacing, which is equivalent to 1-1/2 ordinary single spaces. Where copy has a great many subscripts and superscripts, however, double spacing should be used.

Special Characters may be filled in carefully freehand, using dense black ink, or INSTANT ("rub-on") LETTERING may be used. AMS has a sheet of several hundred most-used symbols and letters which may be purchased for $5.

Diagrams may be drawn in black ink either directly on the model sheet, or on a separate sheet and pasted with rubber cement into spaces left for them in the text. Ballpoint pen is *not* acceptable.

Page Headings (Running Heads) should be centered, in CAPITAL LETTERS (preferably), at the top of the page – just above the blue line and touching it.

LEFT-hand, EVEN-numbered pages should be headed with the AUTHOR'S NAME;

RIGHT-hand, ODD-numbered pages should be headed with the TITLE of the paper (in shortened form if necessary).

Exceptions: PAGE 1 and any other page that carries a display title require NO RUNNING HEADS.

Page Numbers should be at the top of the page, on the same line with the running heads.

LEFT-hand, EVEN numbers – flush with left margin;

RIGHT-hand, ODD numbers – flush with right margin.

Exceptions: PAGE 1 and any other page that carries a display title should have page number, centered below the text, on blue line provided.

FRONT MATTER PAGES should be numbered with Roman numerals (lower case), positioned below text in same manner as described above.

MEMOIRS FORMAT

It is suggested that the material be arranged in pages as indicated below.
Note: Starred items (*) are requirements of publication.

Front Matter (first pages in book, preceding main body of text).

Page i – *Title, *Author's name.

Page iii – Table of contents.

Page iv – *Abstract (at least 1 sentence and at most 300 words).

*1980 Mathematics Subject Classifications represent the primary and secondary subjects of the paper. For the classification scheme, see Annual Subject Indexes of MATHEMATICAL REVIEWS beginning in December 1978.

Key words and phrases, if desired. (A list which covers the content of the paper adequately enough to be useful for an information retrieval system.)

Page v, etc. – Preface, introduction, or any other matter not belonging in body of text.

Page 1 – Chapter Title (dropped 1 inch from top line, and centered).

Beginning of Text.

Footnotes: *Received by the editor date.
Support information – grants, credits, etc.

Last Page (at bottom) – Author's affiliation.

ABCDEFGHIJ – AMS – 898765